What Einstein Did Not See

For: Joseph!
Enjoy! Bill
Tom

What Einstein Did Not See

Redefining Time to Understand Space

Thomas W. Sills

Dearborn Resources · Chicago · 2008

Published by
Dearborn Resources
P. O. Box 59677
Chicago, IL 60659

timespace4d@aol.com

What Einstein Did Not See
PRE-PUBLICATION COPY

NOT FOR DISTRIBUTION TO THE PUBLIC
PUBLICATION DATE: MARCH 2008

FIRST EDITION   ISBN 0-9644096-3-1

Cover design by Dickinson Associates - Ken Spiegel

Table of Contents

CHAPTER 1:
Current Problems with Concepts   **1**

CHAPTER 2:
Is Physical Space Three or Four-Dimensional?   **10**

CHAPTER 3:
Timespace, A New Concept for the Fourth Dimension   **25**

CHAPTER 4:
Is Timekeeping Frequency Universal Time?   **32**

CHAPTER 5:
Universal Time in Four-Dimensional Physical Space   **38**

CHAPTER 6:
Lorentz-FitzGerald Contraction of Timespace Is Time Dilation   **49**

CHAPTER 7:
What Does Four-Dimensional Space Look Like?   **57**

CHAPTER 8:
Time Travel, Time Reversal, and Astronomy   **68**

CHAPTER 9:
Comparison of Special Relativity with Timespace   **78**

CHAPTER 10:
Comparison with General Relativity   **88**

REFERENCES   **97**

INDEX   **134**

Dedicated to
Students of Science Everywhere

Permissions

Cover Photo	Albert Einstein, Used by permission. ©Yousuf Karsh - Woodfin Camp and Associates
Page 1	*As Time Goes By*, by Herman Hupfeld, ©1931 (Renewed) WARNER BROS. INC. All Rights Reserved. Used by permission.
Page 2 Figure 1.1	Albert Einstein, Used by permission. © Yousuf Karsh - Woodfin Camp and Associates
Page 14 Figure 2.4	1881 Natural Philosophy Title Page Thomas W. Sills Library
Page 19 Figure 2.5 Figure 2.6	Herman Minkowski. 1909 *Raum und Zeit* title page. Thomas W. Sills Library
Page 57	*Esau I Saw Esau*, Edited by Iona & Peter Opie, Illustrated by Maurice Sendak, ©1992, 1947 Iona Opie (London: Walker Books) Used by permission.
Page 59 Figure 7.1	Albert Einstein, Used by permission. © Yousuf Karsh - Woodfin Camp and Associates
Page 63 Figure 7.6	Zoomtool Hypercube, Zoome System and Zometool are registered trademarks of Zoomtool, Inc. Zome System used with permission from Zoomtool, Inc. www.zomesystem.com
Page 90 Figure 10.1	Abram Pias Handwriting. Title Page of The Tao of Physics. Thomas W. Sills Library, Jeff Weber Rare Books

Preface

This book reflects my experience as a physics educator of many years. A physics educator studies methods and research on the teaching of physics. Often a physics educator works as a teacher trainer or as an editorial consultant on physics textbook development. The focus of this work is a clear presentation of science concepts for the beginning student.

Physics before the year 1850 bore the name "natural philosophy." This book deeply examines the natural philosophy behind the two most complex concepts in science: Einstein's relativity and Minkowski's four-dimensional space. Both attempt to describe space and time.

This book is a careful walk away from three-dimensional space where readers are so comfortable. Upon leaving, readers enter a reasonable Euclidean view of four-dimensional space that exists around them. Understanding space is possible with the new concepts of Timespace and Universal Time. With new insight into four-dimensional space, readers can proceed to understand higher dimensions of space.

I presented early ideas on this subject March 4 and 11, 2000, at Adler Planetarium in Chicago in a lecture series titled, "Thinking in Four Dimensions to Understand Einstein Relativity."

Thomas W. Sills
October 2007

Acknowledgments

For over ten years, friends, colleagues and acquaintances helped, each in their own way. In 1997, I verbally walked John D. Arthurs out of three-dimensional space into four-dimensional space. At that time John worked to market my previous books. He was first to insist that I write this book.

When I discuss four-dimensional space, my classes consistently become pin-drop quiet. The writing of this book received sabbatical support in spring 2007 from Wilbur Wright College, One of the City Colleges of Chicago.

Francisco Aguilera, James Hawes, Lloyd Hettiger, Corinne Jembrzycki, Joseph Mayer and Walter Trentadue provided valuable manuscript evaluation. Wayne Johnson and Francisco Tapia helped with the photography and the reference library.

In their own way, several individuals told me not to give up. They include Mordecai Benharhari, Bruce Gans, Paul Hewitt, Betty Johnson, Benito Kalaw, Iona Opie, M. Wilson, and Bill Winchief.

Many offered forms of encouragement, energy and support for over ten years. Listing each acknowledgment in this brief account would not be possible. Thanks everyone.

I must admit that after 40 years as a professor, I didn't fully understand the paradoxes of Einstein's relativity until now. I sincerely appreciate those who helped me on this journey.

Chapter 1
Current Problems with Concepts

"With speed and new invention, and things like fourth dimension,
Yet we get a trifle weary, with Mister Einstein's theory,
So we must get down to Earth, at times relax, relieve the tension."

--- *As Time Goes By*, Warner Bros. Music
(With Permission)

PROBLEM ONE:
Einstein's Time Dilation Often Confused with Accelerated Timekeeping
Einstein's equation for time dilation in his 1905 special theory of relativity states timekeeping slows down on clocks moving relative to a clock at rest. Time passes at a slower rate. Einstein's time dilation is Equation 1.1:

$$t = \frac{t_o}{\sqrt{1 - \frac{v^2}{c^2}}}$$ **Equation 1.1**

The symbol t is the tick size on the moving clock. The symbol t_o is the tick size on a clock at rest with the observer. The symbol c is the speed of light.

If an object moves through space at speed v, the bottom of the equation becomes smaller than one. This makes the size of t larger than the size of t_o. With larger size ticks on the moving clock the rate of timekeeping slows down. Yet a person in motion with the moving clock, according to Einstein relativity, will experience timekeeping in a normal way. The person in motion experiences timekeeping that the person would experience at rest! This is the paradox that Einstein's relativity creates.

Students introduced to special relativity for the first time often erroneously conclude that time dilation for moving objects speeds up timekeeping. They think t in Einstein's time dilation formula refers to the rate of timekeeping. Instead, t measures the observed "tick size" of a moving clock. The use of the term *time dilation* also adds to the confusion. This student error seems unimportant. Yet this error is reasonable to the beginning student. Understanding the problem behind this reasonable error is crucial to simplifying Einstein's relativity.

PROBLEM TWO:
The Clock or Twin Paradox Creates Conflicting Realities
　　Time dilation presents the paradox of simultaneous timekeeping systems in conflict. We ask students of Einstein's special relativity to accept the conflicting timekeeping systems. Here is the classic example of the twin trip that so vexes students new to physics. Twins Albert and Alvin are born at the same time. Alvin takes a very fast trip in a spaceship that eventually returns to Earth. Alvin, the traveling twin, observes and measures a natural progression of time. Simultaneously, Albert on Earth also observes and

measures a natural progression of time. Yet Albert's clock measures a larger amount of time than Alvin's clock! Upon Alvin's return he meets his Earth-bound twin Albert. Based upon the length of the trip and the speed of the spaceship Albert can be 30 years older than Alvin!

Albert　　　　　**Alvin**

Figure 1.1

　　Yourgrau, in his book *A World Without Time,* describes the philosophical problem when time is both a measure for motion and a spatial dimension (Yourgrau 2005, 119). Yourgrau reminds us that Gödel, at Einstein's seventieth birthday, presented the philosophical necessity for "t" and "T." Gödel's t is relativistic time and Gödel's T is time with mathematical truth. In 1949 these philosophical considerations were mostly ignored. Today there is a demand for the ideas of relativity to be philosophically lucid.

Chapter Three introduces a new concept, Timespace τ. Chapter Five introduces a new concept, Universal Time T. In Newtonian physics "t" is the time referred to as absolute time. Here we will use the symbol "T" as Universal Time to measure motion along each of four dimensions of space. Universal Time T is similar, but not identical to traditional Newtonian time t. Physics historically uses traditional time t as a measure of motion along each of three dimensions of physical space. With Timespace as a new idea for the fourth dimension, physics can lucidly describe physical space as four-dimensional.

PROBLEM THREE:
Who's Moving?

"Who's moving?" "Who's on first?" Abbott and Costello's classic comedic confusion applies! We ask, "can we consider the traveling twin's ship as the stationary world? Does Alvin observe the Earth move away and then return? Who's moving?" A current popular explanation is that Alvin, the traveling twin, accelerated and decelerated with his trip out and back, (Kaku 2004, *Discover*, 19). The force of acceleration on Alvin makes him the traveling twin.

Yet it seems as reasonable that Alvin could be the stationary twin who sees Albert accelerate away and decelerate as Albert turns around for the return trip. Alvin's spaceship has the rocket engine that makes the trip possible. Alvin feels the force of acceleration and deceleration. Albert on Earth feels no force. Why would this experience of accelerating force make one's clock slow down? We will learn that four-dimensional space is mysteriously different than three-dimensional space.

PROBLEM FOUR:
Consistently Measured Light Speed Makes Length of Moving Objects Contract

Einstein made one bold statement in the special theory of relativity. He stated measurement of the speed of light always has the same value. Today experimental verification exists. In 1905, conclusive experimental verifications were not available to Einstein.

The speed of light will always have the same measured speed regardless of the speed of the light's source. This consistency is not like one dollar always equals 10 dimes. The implication is a laser light beam will not move faster than the speed of light even if the laser is on a fast-moving spaceship. A spaceship moving at half the speed of light fires a laser beam. Those in the spaceship and those outside the spaceship measure the light beam moving at the speed of light. Those outside the spaceship will not measure the beam to move with 50% more speed.

The solution to this strange problem is that space contracts as objects move. Why would the contraction of space make light move at the same speed? An object's length contracts with relative motion. Chapter Six discusses this contraction of space.

PROBLEM FIVE:
How Can the Traveling Twin, Reunite with the Earth Twin?

Alvin measures and experiences a smaller amount of time on the trip away from Earth and back. Yet Alvin observes time progression in a normal way, as if at rest. This is the paradox of Einstein's time dilation. Then how can Alvin have enough time to make the trip?

Wilson in *College Physics* offers a solution to this question (Wilson 1990, 780). Wilson also considers the relativistic contraction of length for the trip. Wilson justifies this conflicting, confusing concept of time measurement with the confusing concept of length contraction. The distance of the total trip for the traveling twin is shorter than that measured by the stationary Earth twin. Why do we need Wilson's justification? So that the twins can reunite.

In the physics of Einstein's relativity, conflicting and confusing ideas justify other ideas. The result is paradoxes seem acceptable.

PROBLEM SIX:
The Triplet Paradox
    Triplets Albert, Betty, and Cathy are born at the same time. Consider a triplet trip paradox where Albert on Earth observes Betty and Cathy accelerate and travel away together at a speed of .8c. Then Betty observes Cathy accelerate and travel away at a speed of .8c. How does Albert observe the timekeeping of Cathy? Does Albert observe Cathy's timekeeping as the addition of Betty and Cathy's velocities? Then Cathy's relative motion at 1.6c exceeds the speed of light. This cannot be possible according to Einstein's special theory of relativity! Exceeding the speed of light is impossible.
    Cathy's timekeeping dilates twice. The formula for the correct velocity of this confusing velocity addition is not simple. What is twice dilated? Equation 1.2 presents the formula (Salmon 1975, 90):

$$V = \frac{u+v}{1+uv/c^2} \qquad \textbf{Equation 1.2}$$

V is the relative speed of Cathy and Albert, u is the relative speed of Betty and Albert, and v is the relative speed of Cathy and Betty. So $V = 1.6c/1.64 = .98$ C. Cathy's speed is less than the speed of light. The triplet paradox again illustrates the complexity of Einstein relativity concepts.
    Einstein's first paper on relativity appeared in 1905 in *Annalen der Physik*, volume 17 (Einstein 1905, 906). The paper was *Zur Elektrodynamik bewegter Körper*. English translation: On the Electrodynamics of Moving Bodies. Physicists identify this paper as the special theory of relativity. Einstein included Equation 1.2 for addition of speeds in his famous 1905 paper. English translations of this 1905 paper are in books by both Miller (Miller 1977 1981, 370) and Stachel (Stachel 1998, 123).

PROBLEM SEVEN:
Minkowski Light Cone Description of Space is Not Clear
    The next chapter discusses how Hermann Minkowski's 1908 paper *Raum and Zeit* combined space and time to produce four-dimensional space. In this paper, he introduced the light cone that you often see with explanations of Einstein Relativity (Taylor and Wheeler 1992, 181). See Figure 1.2:

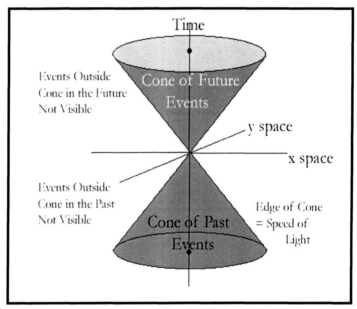

Figure 1.2 (G064f)
Minkowski Light Cone

According to Minkowski, time is the fourth dimension of his idea of a space-time continuum. He also recognizes, as did Einstein, that no measured relative motion can exceed the speed of light. These two ideas allow the volume of a light cone to represent past and present events available to an observer. In Minkowski's new concept of a space-time, events that occur outside this cone are not events available to the observer. That is, light from these invisible events never reach the observer.
    Recall the philosophical question: "How do you know the refrigerator

light turns off when you close the refrigerator door?" According to Minkowski, space-time can best be described with four dimensions: x, y, z, and t. So, events outside the light cone must occur with no observer's knowledge. Perhaps one can just wait for the light's information to arrive with the event information. This solution appears to be a good idea. Yet we all move along the dimension of time. Minkowski's four-dimensional space will not allow this solution. We will not be at the right spot at the right time.

PROBLEM EIGHT:
Terms Spacelike and Timelike are Hard to Accept

The popular *Spacetime Physics* by Taylor and Wheeler provides a helpful introduction to Einstein's special relativity (Taylor and Wheeler 1963, 11 and 172) (Taylor and Wheeler 1992). This book attempts to make Einstein relativity's non-Euclidean geometry graphically visual. This book uses special terms: "Spacelike" and "Timelike" intervals to understand four-dimensional distances between events. Eddington also used these terms in 1920 (Eddington 1920 1959, 60). Minkowski originated these terms in 1908 (Minkowski 1909, 8).

An interval between two positions in four-dimensional, Non-Euclidean space can be either like measurement of xyz S-space or measurement of time. When using spacelike and timelike concepts understanding four-dimensional space can be both easier and extra challenging. How can non-Euclidean space be both spacelike and timelike? In our language space and time are different ideas. Taylor and Wheeler use even more exotic terms such as "Lightlike" and "World Line" to describe that time is not the same in reference frames moving relative to each other. (Taylor and Wheeler 1992, 172)

PROBLEM NINE:

Relative Timekeeping is More Confusing than Newtonian Absolute Time

To better understand the confusion with the twin trip, ask the question: "What time is it?" Newtonian, pre-Einstein physics, only provides one answer. Newtonian time is absolute.

In 1905, Einstein postulated that time varies relative to the clock's observer. In 1920, Eddington clearly described Einstein's contribution to ideas on space and time (Eddington 1920 1959, 211):

> "Both Larmor and Lorentz had introduced a "local time" for the moving system. It was clear that for many phenomena this local time would replace the 'real' time; but it was not suggested that the observer in the moving system would be deceived into thinking that it was the real time. Einstein, in 1905 founded the modern principle of relativity by postulating that this local time was *the time* for the moving observer; no real or absolute time existed, but only the local times, different for different observers."

Einstein's time dilation in his special theory of relativity is confusing. In the twin paradox, an observer measures time at a slower pace if the clock is in relative motion. Yet all observers identically measure time if the clock is at rest with them. This timekeeping paradox is not our common notion of time.

How all clocks clearly agree with each other is the subject of Universal Time "T" presented in Chapter Five. Do not confuse this new universal concept for time with previous uses of "proper time." Proper time is a well-known concept of Einstein's relativity (Taylor and Wheeler 1992, 10). Proper time assumes that the passage of time is the same on all clocks. Proper time is locally observed time on clocks in three-dimensional physical space. Sometimes authors call proper time local time (Taylor and Wheeler 1992, 307), lab time (Taylor, E. And Wheeler 1992, 173), or wristwatch time (Wheeler 1990, 303). Or, in Einstein's German, "Eigenzeit" means "own-time." (Taylor and Wheeler 1992, 11)

1.1 The Core of Each of These Problems is Time and Space Definitions

Readers familiar with Einstein relativity may not find these problems troublesome. Accepting paradoxes about time and space is common for those who "understand" relativity.

At the core of each of the problems presented here is the confusing relationship of time and space. Is time a dimension of space? Does the dimension of time give space four dimensions? Is Minkowski's space-time continuum a clear understanding of the relationship of time and space?

Or, must we now redefine time, as some suggest? As recently as 2006, authors concluded that the real question is the definition of time. Recently distinguished authors have stated the need for time, not Einstein's space-time, to be redefined. In May 2006, professor David Gross, Nobel Laureate, concluded the 2006 Heilborn Distinguished Lectures at Northwestern University by stating: "We may be required to redefine time." Likewise, in the 2006 book, *The Trouble with Physics,* Dr. Lee Smolin stated that time is terribly hard to represent (Smolin 2006, 257):

> "We have to find a way to *unfreeze* time without turning it into space. I have no idea how to do this. ... It's terribly hard to represent time, and that's why there's a good chance that this representation is the missing piece."

1.2 Creative Solution to Einstein Relativity's Confusing Concepts

All of these problems need a new, different solution. This book shows how such a solution comes from modification and clarification of the time concept. We will see the simplicity of examining new ideas for time components. These components are Timespace and Universal Time. These new ideas reveal a simple Euclidean solution for the problem of time and space.

Chapter 2
Is Physical Space Three or Four-Dimensional?

"From now on, space itself and time itself should descend into a shadow and only a union of both should retain its independence."

--- Hermann Minkowski, *Raum und Zeit*, 1908

2.1 What Is a Dimension of Physical Space?

The study of space uses mathematics called geometry. Vectors are lines which connect any two positions in space. Common experiences, like vision and motion, use three-dimensional Euclidean geometry. A minimum of three dimensions defines, or locates, a position in physical space. Physical space is space where motion and the laws of physics apply. Figure 2.1 shows that Cartesian coordinates x, y, and z are three distance measures, each in meters at 90 degrees. Figure 2.2 shows spherical coordinates r, θ, and φ with a measure of r in meters and with measures of θ and φ in degrees. Figure 2.3 shows another three-dimensional system in cylindrical coordinates r, θ, and h. As we see in Figures 2.2 and 2.3 coordinates of some dimensions are angular measurements:

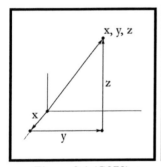

Figure 2.1 (G070)
Cartesian
Coordinates

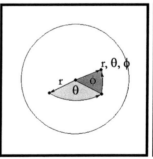

Figure 2.2 (G069f)
Spherical
Coordinates

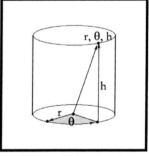

Figure 2.3 (G068f)
Cylindrical
Coordinates

Physical space is where all physics phenomena reside. Before Einstein's theory of relativity, Newtonian motion was the foundation of physics. Motion was fundamental to all ideas in physics. Most fundamental was the measurement of time. With the measurement of time, other ideas evolved. For example, the dynamics that cause a material object's motion. Another is molecular motion as heat. Related ideas include energy, electricity, magnetism, light, and nuclear energy. Physical space is any multidimensional space where motion occurs.

Einstein's four-dimensional space uses non-Euclidean geometry. In this geometry, coordinates of x, y, z, and t describe space. Here we measure space with three dimensions of distance and one dimension of time.

Notice that three-dimensional, Euclidean geometry also measures space in different ways. Spherical and cylindrical coordinate systems use measures of distance and measures of angles. The time dimension in four-dimensional space creates a special problem. Motion is the concurrent measurement of distance and time. If time becomes a dimension of space, then time is not available to describe motion. When time is a measure of motion as well as a dimension of space, then conflicting ideas in Minkowski's four-dimensional geometry result. **Minkowski creates and then masks these problems with the idea that four-dimensional space is a space-time continuum.**

Later chapters propose new ideas. These ideas use Euclidean geometry and motion in all four dimensions. For consistency we require that each dimension of four-dimensional physical space possess the same qualities. These new ideas for physical space distinctly contrast with Minkowski's non-Euclidean ideas. Time and motion will be at the foundation of each dimension in our new four-dimensional, Euclidean physical space.

2.2 Chronological Review of Science Literature on Space and Time

Science authors of countless books and articles on space and time wrote both before and after Albert Einstein's 1905 special theory of relativity. In this section, we mention interesting ideas about space and time in chronological order. We do not imply that these ideas are comprehensive, nor even valid. The selected ideas apply to new concepts in this book and deserve contemplation.

The Leibniz-Clarke correspondence in 1717 on the validity of Newtonian absolute space and time provided a temporary emancipation for the natural sciences from philosophy and theology (Alexander 1956) (Jammer 1954 1957, 117). This emancipation occurred over the two succeeding centuries before Einstein's 1905 special theory of relativity. In one correspondence in 1716, Dr. Clarke argues that "space is not a substance, but a property." (Alexander 1956, 47)

In 1905, in the electrodynamic part of the special theory of relativity, Albert Einstein describes "the Maxwell-Hertz equations for empty space." See English translation in *Special Theory of Relativity* by Arthur Miller (Miller 1981, 404). But, near the end of his career, Albert Einstein emphasizes that empty space does not exit. See Chapter Ten for a discussion of these contradictory statements about empty space.

In 1913, H. Poincaré authored *The Foundations of Science*. Here this great mathematician makes the following profound statement (Poincare 1913 1946, 428):

"We see that if geometry is not an experimental science; it is a science born apropos of experience; that we have created the space it studies, but adapting it to the world wherein we live. We have selected the most convenient space, but experience has guided our choice; as this choice has been unconscious, we think it has been imposed upon us..."

In 1939, in *The Philosophy of Physical Science*, Sir Arthur Eddington states that in the physical universe, matter occupies only a small region compared with empty space (Eddington 1939 1940, 184). In this same book, Eddington stresses the importance that both the experimenter and the theorist use the same language (Eddington 1939 1940, 72).

In 1954, in his classic book, *Concepts of Space: The History of Theories of Space in Physics*, Max Jammer concludes that like all science, the science of space is still unfinished business (Jammer 1954 1957, 190).

In 2001, Sir Martin Rees, Royal Astronomer in Great Britain, concludes in his book *Our Cosmic Habitat*, that the space in our universe must be three-dimensional. Sir Martin Rees reasons that if there were a fourth spatial dimension, the area of a sphere would be proportional to r^3 instead of r^2. So, the gravitational force of attraction would follow an inverse cube law. He concludes that things would be catastrophically different if gravity obeyed an inverse cube law. A planet that slightly slowed down would plunge into the Sun (Rees 2001, 148).

Again in 2001, in contrast with this same book, Sir Martin Rees accepts the idea that space has more than three dimensions (Rees 2001, 149):

"Currently nothing seems absurd about a universe where space has extra dimensions: according to superstring theories, the ultra-early universe had ten or eleven."

Brian Greene is a proponent of superstring theory and higher-dimensional spaces. In 2004, in *The Fabric of the Cosmos* Greene concludes that there is a major gap in our understanding (Greene 2004, 486):

"Or, perhaps I should say, when it comes to identifying spacetime's elemental ingredients, we have no idea about which we're really confident. This is a major gap in our understanding . . . "

In 2005, three published books by different authors defined higher dimensions of space used by scientists. Each book includes an extensive glossary. None of these books present a definition for space (Krauss 2005, 257) (Kaku 2005, 381) (Randall 2005, 459).

In 2006, Lee Smolin in *The Trouble with Physics* concludes that we need to find a way to unfreeze time. He states that he has no idea how to do this. He reasons that it is terribly hard to represent time. For Smolin a new representation of time is the missing piece in physics needing discovery (Smolin 2006, 257).

In 2006, Max Jammer concludes in *Concepts of Simultaneity* that time differs from space (Jammer 2006, 300).

2.3 Chronological Review of Philosophy Literature on Space and Time

Physics is "natural philosophy." In fact, authors title physics textbooks prior to the mid-nineteenth century *Natural Philosophy.* (See www.aip.org/history/newsletter/spring2004/physics-texts.htm). Motion in space is at the foundation of Newtonian physics. Physics implies physical space and natural philosophy. Figure 2.4 is a title page from a 1881 physics textbook:

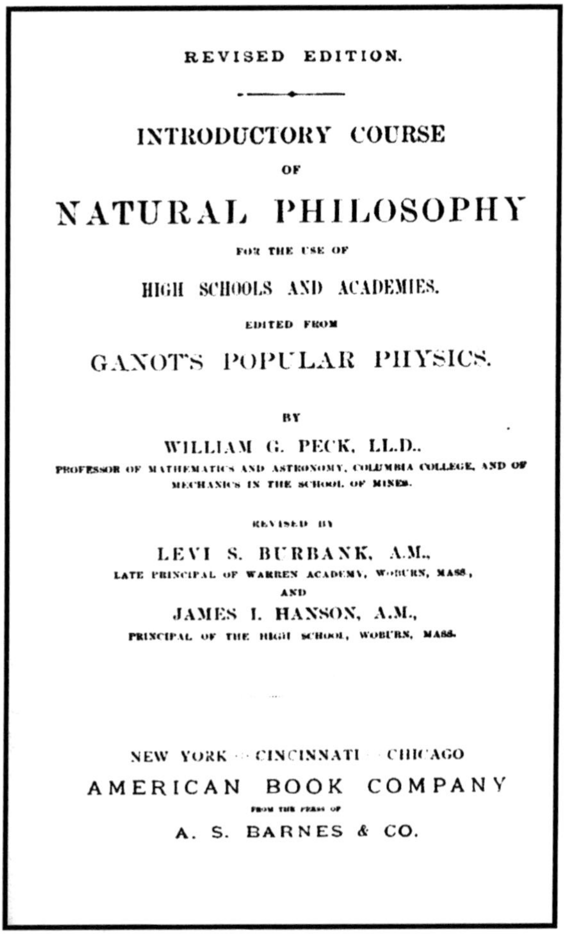

REVISED EDITION.

INTRODUCTORY COURSE

OF

NATURAL PHILOSOPHY

FOR THE USE OF

HIGH SCHOOLS AND ACADEMIES.

EDITED FROM

GANOT'S POPULAR PHYSICS.

BY

WILLIAM G. PECK, LL.D.,

PROFESSOR OF MATHEMATICS AND ASTRONOMY, COLUMBIA COLLEGE, AND OF MECHANICS IN THE SCHOOL OF MINES.

REVISED BY

LEVI S. BURBANK, A.M.,

LATE PRINCIPAL OF WARREN ACADEMY, WOBURN, MASS,

AND

JAMES I. HANSON, A.M.,

PRINCIPAL OF THE HIGH SCHOOL, WOBURN, MASS.

NEW YORK ·· CINCINNATI ·· CHICAGO

AMERICAN BOOK COMPANY

FROM THE PRESS OF

A. S. BARNES & CO.

Figure 2.4 (G071f)
Title Page of 1881 Physics Textbook

Authors in the field of pure philosophy also wrote books and articles about space and time, before and after the special theory of relativity in 1905. In this section, we again mention ideas in chronological order which are interesting. These ideas also apply to the new ideas in this book and deserve contemplation.

One philosopher, C. D. Broad suggests that since quantities for speed are relative, physics as a philosophy is not valid. In 1914 C. D. Broad in *Perception, Physics, and Reality* makes a philosophical argument critical of physics and Newtonian use of absolute values for motion and time (Broad 1914, 277):

"If you take all motion to be relative, you cannot talk about THE direction or THE velocity of a body, because no body has a unique direction or velocity. Each one will have dozens of different and equally good velocities and directions according to the bodies relatively to which its motion is determined. Similarly for the measurement of time. It is always measured by the recurrence of certain events, and, according as you take different processes to be uniform, you will reach different results as to the uniformity or lack of it in a given motion. Newton of course clearly recognized this. He distinguished between absolute space, time, and motion, and the relative spaces, times, and motions; and admitted that it was perfectly possible that no change was uniform and therefore capable of giving a measure of absolute time, and that no body was absolutely at rest or in uniform absolute motion in a straight line. But he thought that he could give a criterion of absolute motion, and, by experiment, prove the reality of at any rate absolute rotation. On the other hand, most philosophers have a strong objection to absolute motion, space, or time; and putting this aside, there is felt to be a reasonable doubt whether it can be necessary or possible to start--as mechanics presumably does--from empirical data, to reach laws which are stated in terms of two entities which certainly cannot be perceived, and finally to return and explain empirically perceived motion by means of these laws."

In 1920, philosopher S. Alexander said that what is contemplated as physical space-time is enjoyed as mental space-time (Alexander 1920 1950, 180). Alexander concludes that total space-time involves two elements, total space and total time (Alexander 1920 1950, 86).

In 1920, Norman Campbell in *Foundations of Science* concludes that absolute magnitude has no physical significance. His reasoning is that if one reconstructs the whole universe on a different scale, no experiments could detect the change (Campbell 1920 1957, 416).

In 1925, Edwin Burtt in *The Metaphysical Foundations of Modern Physical Science* concludes that great scientists in history should be profound philosophers as well as acute scientists. If they were, they could hardly remain satisfied with an answer that failed to probe the science to its very depth (Burtt 1924 1954, 325).

In 1927, A. D'Abro authored *The Evolution of Scientific Thought*. In this comparative study D'Abro reminds us physicists speak of absolutes and relatives. When physicists use these terms, they are referring to categories differing from those of philosophers. As far as scientific philosophy is concerned, absolute reality is a myth. In the final analysis nature reduces to structure, that is, to relationships (D'Abro 1927 1950, 99). D'Abro states that space-time is a fundamental continuum. Separately space and time, varying as they do with the observer's motion, have no absolute significance (D'Abro 1927 1950, 279).

In 1931, similar philosophical questions agreed with C. D. Broad's 1914 critique of physics. See J. Callahan's *Euclid or Einstein: A Proof of the Parallel Theory and A Critique of Metageometry* (Callahan 1931).

In 1931, L. L. Whyte concludes in *Critique of Physics* that because Einstein's theory uses a four-dimensional continuum with an imaginary time-coordinate, Einstein's theory is not a general or final theory of physical space and time (White 1931, 179). See Equation 2.1 and Equation 2.2 later in this chapter. Whyte also states that the Einstein theory's interpretation of the invariant c as a velocity and the demand for an invariant metrical representation of the observed topological facts are incompatible (White 1931, 45). In other words, the speed of light c is the invariant, or the constant. The invariant is not the c times t distance that light travels. The distance that light travels is a metric, called ct.

In 1952, C. D. Broad in *Ethics and the History of Philosophy* makes the interesting statement (Broad 1952, 186):

"It seems to me then that there is no close logical connection between the controversy about Absolute and Relative Space, on the one hand, and these controversies about the finite and infinite extent of the universe, and the existence or non-existence of empty spaces within it, on the other."

In 1953, G. E. Moore in *Some Main Problems of Philosophy* made the observation that time is and exists as a fact, nevertheless time is not real (Moore 1953, 214).

In May 1955, G. J. Whitrow in *Why Physical Space Has Three Dimensions* concludes that the nature of space is three-dimensional. G. J. Whitrow concludes that the problem is not solved (Whitrow 1955, 13):

"Since the mathematical discovery of higher space, a clear-cut problem has emerged concerning the origin of the three-dimensional character of physical space. Despite various recent attempts to show that this feature is either a necessary attribute of our conception of physical space or is partly conventional and partly contingent, the problem cannot be considered as finally solved."

In 1957, Phillipp Frank in *Philosophy of Science* makes the following statement about four-dimensional space (Frank 1957 1974, 162):

"The four-dimensional formulation of relativity is a useful instrument for the presentation of physical events, but it cannot be interpreted in our everyday language by simply speaking about the four-dimensional space-time continuum as we have been accustomed to speak about our ordinary three-dimensional space."

In 1966, Bas van Fraassen in *An Introduction to the Philosophy of Time and Space* concludes that space-time is a mathematical structure (Van Fraaseen 1970, 198).

In 1975, Wesley Salmon in *Space, Time and Motion* compared pure geometry to applied geometry. Pure geometry is a system of pure mathematics. Applied geometry is a synthetic geometry physicists use to describe the physical world. According to Salman pure mathematics will never yield a description of the physical world (Salmon 1975, 28).

On the topic of space and time, philosophical discussions seem to be endless. One common thread to all of these presentations is that either we need better ideas or space has only three dimensions.

2.4 Conclusions: Space Has Three Dimensions, or We Need New Ideas

Scholars of science and philosophy arrive at conflicting conclusions. Some conclude that three-dimensional space is the best way to describe space. Others consider space to have four-dimensions or more. Many conclude that we need new ideas. A consensus among authors on the nature of space does not exist.

2.5 Four Dimensional Space and Motion Before Einstein Relativity

In the nineteenth century public interest in dimensions of space became popular. In 1880, Edwin A. Abbott wrote *Flatland* (Abbott and Burger 1880 1994). This popular book describes a strange world where space is two-dimensional. In 1960 after Einstein's relativity theory, the natural sequel to Flatland was *Sphereland* by Dionys Burger (Abbott and Burger 1880 1994). This book compared the universe to the surface of a sphere.

In 1904, one year before Einstein's 1905 publication on special relativity, C. H. Hinton wrote a book titled, *The Fourth Dimension* (Hinton 1904). In a July 1904 Harper's Monthly Magazine article, *The Fourth Dimension*, Hinton considers motion in the fourth dimension (Hinton 1904, 233):

> "It is in the examination of questions such as these that the physical inquiry as to the existence of the fourth dimension consists in asking, namely, whether the types of action which occur can be explained on the principles of three-dimensional mechanics, or whether they demand for their explanation the assumption of a four-dimensional motion."

In 1905 Poincaré studied four-dimensional space (White 1931, 175) (Poincaré 1913 1952, 93). Poincaré believed that one could use either Euclidean or non-Euclidean geometry as the geometry of space. He argued that Euclidean three-dimensional space would be the preferred space used by physicists. One hundred years later we find overwhelming experimental verification for four-dimensional space and Einstein's theory of relativity.

2.6 Minkowski Added the Fourth Dimension to Einstein's Relativity

In September 1908, Hermann Minkowski in Figure 2.5 presented his four-dimensional solution to special relativity (Minkowski 1909). In 1909, Minkowski's ideas were published in a booklet titled *Raum und Zeit* (Space and Time). See Figure 2.6. Minkowski's four-dimensional space-time continuum was not developed by Einstein. Time-space is not a coherent development of ideas by one author. In fact, at an earlier time Minkowski was Einstein's university mathematics instructor at the University in Göttingen.

RAUM UND ZEIT

VORTRAG, GEHALTEN AUF DER 80. NATUR-
FORSCHER-VERSAMMLUNG ZU KÖLN
AM 21. SEPTEMBER 1908

VON

HERMANN MINKOWSKI

MIT DEM BILDNIS HERMANN MINKOWSKIS
SOWIE EINEM VORWORT VON A. GUTZMER

LEIPZIG UND BERLIN
DRUCK UND VERLAG VON B. G. TEUBNER
1909

Figure 2.6 (G020)
1909 *Space and Time*
Title Page
Minkowski's Idea for
Four-Dimensional Space

Figure 2.5 (G022)
Hermann Minkowski

In 1854, at the University in Göttingen, Georg Friedrich Bernard Riemann was an early champion of non-Euclidean geometry as a mathematical tool (Carslaw 1906 1912) (Eisenhart 1926). In Einstein's 1905 relativity theory included a Pythagorean equation of invariance. See Einstein's Equation 2.1. Minkowski could see upon rearrangement of Equation 2.1 that three dimensions of space and one dimension of time could describe a four-dimensional, non-Euclidean space. See Minkowski's Equation 2.2:

Einstein $$x^2 + y^2 + z^2 = (ct)^2$$ **Equation 2.1**

Minkowski $$x^2 + y^2 + z^2 + (\sqrt{-1}c)^2 t^2 = 0$$ **Equation 2.2**

In Equation 2.2 x, y, z, and t are the four coordinates of four-dimensional, non-Euclidean space. We can visualize Euclidean space in Figure 2.7. This figure has three components along dimensions x, y, and z:

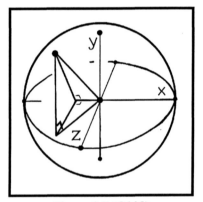

Figure 2.7 (G008)
Euclidean Three-Dimensions

Algebraic expressions best describe the nature of four-dimensional, non-Euclidean geometry. We cannot visualize non-Eulidean geometry as we can visualize three-dimensional Euclidean geometry in Figure 2.7.

This four-dimensional geometry measures space in units of distance for x, y, and z, and in units of seconds, hours and years for time. Minkowski, in 1908, concludes that space and time is a continuum. The following quote is translated into English from *Raum und Zeit* (Infield 1950, 45) (Mlodinow 2001, 192). Space-time is the name Minkowski gives to his continuum of space and time:

> "Gentlemen! The views of time and space that I wish to develop before you grew on the soil of physical experiments. There lies their strength. Their tendency is radical. From now on, space itself and time itself should descend into a shadow and only a union of both should retain its independence."

Perhaps Einstein's creative mind was not always on the academic classroom subject. Perhaps Minkowski enjoyed demonstrating this lack of attention with *Raum und Zeit*. Later, when Einstein finished his general theory of relativity in 1913 (Einstein 1960, 7) (Goldsmith 1997, 95), he became very proficient in non-Euclidean geometry.

Minkowski's fourth dimension is not space like Newton's physical three-dimensional space. Newton founded physics on the idea of motion in physical space. One can see that the fourth dimension created by Minkowski's work is not founded on motion in space. Equation 2.2 shows that Minkowski's fourth dimension is mathematical, not physical.

2.7 Steinmetz Attempts to Clarify Minkowski's Space-time Continuum
In 1923, Charles Proteus Steinmetz in *Four Lectures on Relativity and Space* clarified the equality of Minkowski's four dimensions of the space-time continuum (Steinmetz 1923, 40):

> "In this four-dimensional manifold of Minkowski, this world or time space, which includes symmetrically the space and the time, with x, y, z, u as coördinates, we cannot say that x, y, z are space coördinates and the u the time coördinate, but all four dimensions are given in the same units, centimeters or miles, or, if we wish to use the time unit as measure, seconds; but all four dimensions are symmetrical, and

each contains the space and time conceptions. Thus there is no more reason to consider x, y, z as space coördinates and u as time coördinate than there is to consider x and u as space and y and z as time coördinates, etc."

Steinmetz says that space-time does not imply time is the fourth dimension. Space-time, limited to an algebra equation, represents Minkowski's four-dimensional geometry. Minkowski's creates ideas of timelike and spacelike in an attempt to unite distance measures in xyz S-space and time measures in the fourth dimension.

Is physical space three or four-dimensional? Published speculations do not have a consensus. Experimental verification of four-dimensional space is overwhelming. So, how does this book resolve that the universe must have four dimensions of physical space?

Newtonian, three-dimensional space and Newtonian absolute time cannot explain experimental verifications of Einstein relativity. Yet, Minkowski's four-dimensional space-time continuum fails many who try to understand these ideas. **For example, Minkowski's space-time continuum fails to lucidly explain the twin paradox presented in the previous chapter.**

Although Einstein and Minkowski's approach to time and space is a creative description of physical space, a clear understanding requires something new. The new approach to physical space includes new ideas: *Timespace* and *Universal Time*.

Before we propose new, clearer ideas about space and time, you should first understand the physical space that you use when you think.

2.8 Thinking about Motion with Experience Limited to Three-Dimensional Space

"Who's on first?" and "Who's moving?" In the twin paradox, is the Earth twin or the traveling twin moving? Stop! When asking that question, you assume that space is three-dimensional. Space is not as it seems. In fact, three-dimensional concepts like volume, inside of a room, and outside of a room no longer exist in four-dimensions.

Gedankin experiment: Einstein often performed Gedankin experiments, that is, pure thought experiments. Try a simple thinking experiment that Einstein regularly enjoyed. Ask yourself what is on this page between the two stars?

* 3 *

If you answered "three," you are wrong. No, not a symbol, nor curved lines. So, what is the answer? The answer: Between the two stars there is ink on the page. A silly riddle, you say? No, this experiment shows an important fact. Your hard-working childhood school teacher brainwashed you to recognize this ink shape as the number three.

Likewise, life experiences have brainwashed you to recognize the space around you as three-dimensional. Chapter Eight discusses what happens to a twin on a 747-jet instead of a fast-moving spaceship. This story explains why earthbound creatures only have experience with three-dimensional space.

We should recognize that we have many common notions which may or may not be valid:

- Time is one idea.
- Clocks measure time in seconds, hours and years.
- We can visualize four-dimensional space.
- Four-dimensional space must be similar to three-dimensional space.
- In relativity theory, a space-time interval separates events like a distance interval separates points in three-dimensional space.
- Concepts from three-dimensional space exist in four-dimensional space. Examples:
  - volume and density
  - cube, sphere and pyramid
  - inside and outside
  - center
  - point
  - force and acceleration.

You will hear authoritative scientists with significant knowledge of relativity theory say, "No one knows what happens as you enter the spherical event horizon surrounding a black hole." True, no one has entered and returned from a black hole. Black holes are unbelievable small. One can understand Black holes best with four-dimensional space, not three-dimensional spheres.

2.9 Simplifying Four-Dimensional Physical Space Requires New Concepts

Physical space is not three-dimensional. Yet, Minkowski's space-time continuum is not a simple approach to understanding four-dimensional space. A better understanding that physical space has a four-dimensional structure comes from the creation of new concepts: *Timespace* and *Universal Time*. Together, these new concepts replace and redefine traditional time.

The next chapter introduces *Timespace*. Timespace is a new idea for time. Timespace is a time displacement vector for physical space's fourth dimension.

Chapter 3
Timespace,
A New Concept for the Fourth Physical Dimension

"Time must go somewhere."
--- Anonymous

3.1 Scalar or Vector? Measurement and Classification in Physics

Beginning physics students must sort physics concepts into two major categories. Either the concept is a *scalar* or it is a *vector*. Scalars have no physical direction. Vectors are quantities with an associated direction in physical space. Because they possess only measurable magnitude, or size, mass and energy are easily scalars. Beyond having a measurable magnitude, vector concepts like velocity, force and momentum also include an associated direction in physical space:

◆ a velocity of 30 mph west
◆ a force pulling with 100 pounds
◆ a vertical momentum of 5.6 kg-m/sec

In physics tradition teaches that time is not a vector, but a scalar. Yet time has directional qualities like past, present and future. So, why isn't time a vector? The instructor explains that past, present and future do not have direction in physical space. Instead, these qualities have abstract direction. If time is a scalar, one might conclude that physical space must be three-dimensional. If time is a vector, time must have a physical direction. Time would then become a dimension of physical space. Conventionally time is a scalar. Values for time can be:

◆   15 seconds
◆   10 hours
◆   23.5 years

Introductory students of physics struggle with understanding that time is a scalar. Via logic we see that time is a dimension of physical space as well as a scalar measure that describes motion. How can we give time this quality of a space dimension? The answer is the new concept of **Timespace**.

3.2 Timespace, Vector Displacement in Meters

Saying "eight seconds to the left" makes no sense, or does it? "Eight seconds to the left" might mean that the object moved to the left and it took eight seconds. Taking this one step further, consider giving time the quality of motion and direction. In this case, the direction is along a physical displacement of space. This physical displacement is a vector displacement along the fourth dimension. Call it **Timespace**.

How can this idea for four-dimensional Timespace exist? The fourth dimension of space isn't visible. Physics concepts like velocity, force and momentum exist, and they aren't visible. Yet we can carefully measure these variables in scientific units. We will see that one can also define and measure Timespace.

Timespace is a new concept for time as the fourth physical dimension. Timespace is a vector quantity with physical direction. But with the progression of time don't all objects move with time in the direction of Timespace? According to Einstein's theory of relativity time progresses at varying rates.

3.3 All Objects Move at the Speed of Light in Four Dimensions

Remember the twin trip paradox. Let's apply Timespace to this paradox and see if we can clean up the confusion. The Earth twin ages at the maximum rate. The traveling twin ages slower because the traveling twin accelerated through xyz S-space. We can easily represent the twin trip paradox with Equation 3.1. Figure 3.1 is a visual plot of speed through xyz S-space versus speed through the fourth dimension, Timespace τ:

$$v_s^2 + v_\tau^2 = c^2 \qquad \text{Equation 3.1}$$

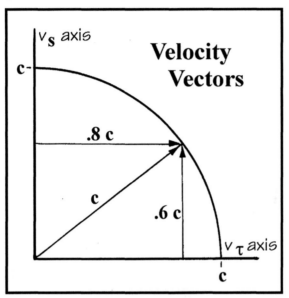

Figure 3.1 (G037A)
Speed in Space vs. Speed in Timespace

The maximum speed toward Timespace, or time displacement, is the speed of light. The maximum speed toward three-dimensional, xyz S-space is also the speed of light. **Motion in xyz S-space reduces the amount of motion in Timespace. The total speed in four-dimensions is always the speed of light.**

Historical Note: Rene Descartes is the father of analytical geometry (Descartes 1677) (Descartes 1925 1954). He was the first person to show how algebraic expressions represent geometry. The Pythagorean Theorem is an example. On the next page, Equation 3.2 represents the geometry in Figure 3.2:

$$a^2 + b^2 = c^2 \qquad \text{Equation 3.2}$$

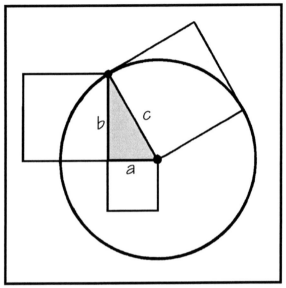

Figure 3.2 (G006)
Pythagorean Triangle

In Rene Descartes' writing, you see the expression "b square" instead of "b squared." His algebra literally envisioned the squares of the Pythagorean Theorem.

Motion along the fourth dimension of Timespace τ is time displacement in meters. Time displacement in meters is a direction in physical space. Time displacement is the quality of the passage of time. The velocity of motion along this fourth dimension of Timespace τ measures the rate of timekeeping, v_τ. A clock at rest, where v_s equals zero, moves with the maximum speed in the fourth dimension of Timespace. This maximum speed v_τ equals c, the speed of light. If a clock moves with the speed of light in xyz S-space, its timekeeping stops because its speed toward Timespace is zero.

The total speed of all objects in four dimensions is c, the speed of light. This fact applies regardless of the object's condition of motion in three-dimensional xyz S-space. See Equation 3.1 and Figure 3.1.

However strange, the principle that the speed of all objects in four dimensions is the speed of light is not new. In 1992 Lewis Carroll Epstein in *Relativity Visualized* (Epstein 1981 1997, 81) and in 2004 Brian Greene in *The Fabric of the Cosmos* (Greene 2004, 49) present this principle. Epstein uses the traditional concept of proper time in his explanation. Greene states that this principle refers only to Einstein's special theory of relativity. **Our objective in this book is to understand much more about this idea and its geometry.**

Figure 3.1 simplifies the representation of four-dimensional space with a two-dimensional diagram. Penrose and Hawking use a two-dimensional diagram of xyz S-space versus time t to represent four-dimensional space (Hawking 1996, 6). Their two-dimensional diagrams suggest that a one-dimensional vector for xyz S-space adequately represents three-dimensional space. S-space and time t are orthogonal and are complete representations of non-visible, four-dimensional space. The complex mathematics of non-Euclidean geometry offers a rich geometry to represent this four-dimensional space. Penrose (Hawking 1996, 105) uses the complex mathematics of twistors. But, with twisters three-dimensional space defines space.

Yet, this new fourth dimension of Timespace is not an independent dimension. Motion in the fourth Timespace dimension is a function of motion in xyz S-space. Equation 3.1 gives the relationship between speed in Timespace, v_t, and speed in xyz S-space, v_s.

3.4 Timespace Simplifies Four-Dimensional Physical Space

Time as a physical dimension isn't a new idea in traditional, three-dimensional physics. In analytical geometry and calculus we plot graphs of distance, velocity, acceleration, and even force versus a horizontal spatial axis of time. Physics analyzes time as a spatial dimension, from introductory physics to advanced engineering physics.

Now, we are formally giving time the quality of passage of time, or direction. Timespace is the fourth physical dimension, where displacement represents the "passage of time." Timespace is displacement that we measure in meters. Timespace displacement can occur at different rates. **Timespace displacement can even be a negative displacement of time reversal!**

Timespace as a vector of time displacement gives the fourth dimension parallels to three-dimensional xyz S-space:

◆ We measure Timespace in units of meters, like displacements in xyz S-space.

◆ Motion occurs along each of the four dimensions.

◆ The maximum speed in any direction of four-dimensional space is c, the speed of light.

◆ Time as Timespace now becomes a vector quantity with physical direction along the fourth dimension.

◆ Past, present and future become measurable as physical displacement. Timespace displacements are like displacements in xyz S-space.

For Earth residents, measuring Timespace in miles is not as convenient as measuring time in hours and years. We will understand this inconvenience in Chapter Seven, *What Does Four-dimensional Space Look Like?* An age of 39 Earth years translates into an age of 230 trillion miles old. Celebrating this birthday could depress anyone.

3.5 Timespace Requires another New Concept for Time

Newton and others founded physics on the concepts of displacement, motion, and time. Newtonian time is absolute. Newtonian time is the same everywhere in three-dimensional xyz S-space. Einstein and Minkowski use non-Euclidean geometry (Wolf 1988, 116). Non-Euclidean geometry replaces the traditional concept of motion with a four-dimensional space-time continuum. A postulate of Einstein's non-Euclidean geometry of space and time is this: All motion in space is relative to the observer. Timespace becomes one of the four dimensions of physical space along with x, y, and z dimensions.

Introducing Timespace τ along with a Universal Time T allows for motion measurement in each of the four dimensions. Using algebra or calculus, Equation 3.2 describes speed in the fourth physical dimension of Timespace:

$$v_\tau = \Delta\tau/\Delta T \qquad \text{or} \qquad v_\tau = d\tau/dT \qquad \text{Equation 3.2}$$

In Equation 3.2 v_τ is the velocity through Timespace τ. This velocity represents the rate of a clock's timekeeping speed.

For Timespace τ to exist as the fourth physical dimension, an absolute, universal measure of time, T, must also exist. This **Universal Time T** must be available when making measurements of motion in each of the four dimensions x, y, z, and Timespace τ. At first one considers measuring a consistent, Universal Time to not be possible. If we observe moving clocks, their timekeeping speed is at a slower rate. Not all clocks will identically measure time. Einstein's time dilation in Equation 1.1 describes this slower rate.

The precedent for Universal Time is Newtonian absolute time that measures motion in only three dimensions. Yet, traditional Newtonian absolute time and Universal Time are different. Universal Time would be a synchronous and isochronic measurement by everyone in the universe. It would be used by everyone to measure motion in each of four dimensions. Such measurements include measurements of motion through Timespace.

If Timespace is a vector displacement along the fourth dimension, does a measure of time with scalar quality exist? The next chapter describes the nature and qualities of frequency as a measurable scalar time. Can time keeping frequency be a universal scalar time measuring motion in each of four dimensions? We will find that examining timekeeping frequency as a measure of scalar time will be enlightening. But, examining this frequency is not fruitful in the search for Universal Time T.

Chapter 4
Is Timekeeping Frequency Universal Time?

"Time is the most undefinable yet paradoxical of things;
the past is gone, the future has not come,
and the present becomes the past even while we attempt to define it."

--- Caleb C. Colton 1780-1822 (Naber 1992 2003, 675)

4.1 Aristotle's Definition of Time, ca. 350 BC

Everyone knows what time is. Putting it into words is another matter. Aristotle was one of the earliest writers who tried to describe time. Traditional physics uses much of this definition of time. The following is Aristotle's definition for time in his own words (*Aristotle's Physics* 1969, 80):

> "Since a thing in motion is moved from something to something else and every magnitude is continuous, a motion follows a magnitude; for a motion is continuous because a magnitude is continuous, and time is continuous because a motion is continuous (for the time elapsed is always thought to be as much as the corresponding motion which took place). Now the prior or the posterior are attributes primarily of a place, and in virtue of position. So since the prior and the posterior exist in magnitudes, they must also exist in motions and be analogous to those in magnitudes; and further, the prior and the posterior exist also in time because time always follows a motion. Now the prior and the posterior exist in a motion whenever a motion exists, but the essence of each of them is distinct [from a motion] and is not a motion. Moreover, we also know the time when we limit a motion by specifying in it a prior and a posterior as its limits; and it is then that we say that time has elapsed, that is, when we perceive the prior and the posterior in a motion."

In this early definition we find that time is no more than the idea of motion.

When students read these words by Aristotle, they find this complex definition ironic. Aristotle made great contributions to the philosophy of logic and reasoning. Aristotle states that knowledge starts with clear definitions. We can't describe the above definition for time as a clear concept, nor as a concept for what clocks measure. Yet, the current physics definition for time is that which clocks measure.

4.2 Searching for the Universal Measurement of Time

Our approach to four-dimensional, physical space requires a universal measure of time. Newtonian absolute time for motion in three-dimensional physical space is like the universal time we need for four-dimensional physical space. Universal Time timekeeping must measure motion in each of the four dimensions of physical space: x, y, z, and Timespace τ.

In traditional physics, if the velocity vector of a car is north at 55 mph on the highway, then the car's speed is a scalar value of 55 mph. Does a similar and simple scalar time exist for Timespace vector displacement? The answer is no! It's just not that simple. Defining Universal Time is the topic of our next chapter.

Timespace τ is the fourth physical dimension, or vector space. We measure Timespace displacement in meters. A universal measure of a scalar time must determine motion. Such measurements of motion would be speed and velocity. Speed and velocity measurements must describe each of the four dimensions of physical space. The new concept for universal scalar time must have no quality of direction, nor quality of the passage of time.

Earlier, Problem One in Chapter One discussed how introductory students find learning the meaning of Einstein's equation for time dilation difficult. Often, after hearing time dilation, students incorrectly associate an increased rate of timekeeping on moving clocks. Einstein's equation of time dilation refers to the length of the observed clock's tick. Actually as the tick grows, the moving clock's timekeeping rate slows down.

Einstein's decision to let t represent the size of the moving clock's tick was a forced choice. **When Einstein presented his theories of relativity, no physical measurement in traditional physics described timekeeping's variable rate.** Traditional time, t, was the only available measurement. Students incorrectly think time dilation increases the clock timekeeping rate for a moving clock.

Together, Timespace and Universal Time replace traditional time. Together Timespace τ, the fourth physical dimension, and Universal Time T allow for the measurement of motion along the fourth dimension and variations in timekeeping associated with speed. The measurement of various speeds in the Timespace dimension is now possible. See Equations 4.1, 4.2 and 4.3:

$$v_s^2 + v_\tau^2 = c^2 \qquad\qquad \textbf{Equation 4.1}$$

$$\textbf{With algebra} \qquad v_\tau = \Delta\tau / \Delta T \qquad\qquad \textbf{Equation 4.2}$$

$$\textbf{Or with calculus} \qquad v_\tau = d\tau / dT \qquad\qquad \textbf{Equation 4.3}$$

Equation 4.1 states that an object's speed in four-dimensional space will always be the speed of light. Adding speed in three-dimensional space v_s to the speed in Timespace v_τ vectorially adds to c, the speed of light, at all times. See Chapter Three Figure 3.1. We discuss these equations throughout this book in detail.

First consider: How do observers measure Universal Time, when each moves at different speeds? If Timespace is vector time, perhaps there exists a scalar concept for time. Perhaps this scalar time is the Universal Time T we need to measure speed in four-dimensional physical space.

4.3 Is Timekeeping Frequency Scalar Time?

Is traditional timekeeping clock frequency f a scalar measure of time? Frequency's measurement units are ticks per second or cycles per second. These units are equivalent to the international unit Hertz (Hz). The frequency of timekeeping is an inherent, scalar property of all clocks.

Frequency has some qualities as a scalar time that we need for the Universal Time T. The frequency of a timekeeping instrument has no physical direction. Nor does frequency have any quality of the passage of time.

Frequency makes some sense as scalar time. Scalar time, as a timekeeping frequency has:

♦ No beginning.
♦ No end.
♦ No direction for the passage of time.
♦ No zero value condition.

4.4 Scalar Time Unites Eastern and Western Philosophies

Fritjof Capra, in the *Tao of Physics*, portrays eastern philosophy's concept of time by quoting Zen master Dogen (Capra 1975 1980, 173):

> "It is believed by most that time passes; in actual fact it stays where it is. This idea of passing time may be called time, but it is an incorrect idea, for since one sees it as passing, one cannot understand that it stays just where it is."

Frequency as a scalar concept for time resonates with eastern philosophy. Time includes the three western philosophy ideas of past, present and future. So many refer to the direction of past or of future as the arrow of time (Coveney and Highfield 1990) (Savitt 1995 1998) (Morris 1985).

A scalar is any measured quantity that has no associated direction in physical space. Although traditional time is a scalar in physics, timekeeping frequency is more a scalar concept. A frequency of 60 cycles per second, or 60 Hertz, stays just where it is. Sixty cycles per second does not move in a direction of north, left, or down.

In their local environment everyone everywhere measures frequency in the same way. How then can we accept frequency of vibration in timekeeping instruments as a time measurement? Einstein states everyone everywhere measures frequency in the same way. On the next page, Einstein's words from the first postulate of his 1905 special relativity theory implies this idea (Miller 1977 1981, 370):

> "..to the concept of absolute rest there correspond no properties of the phenomena, neither in mechanics, nor in electrodynamics, but rather that as has already been shown to quantities of the first order, for every reference system in which the laws of mechanics are valid, the laws of electrodynamics and optics are also valid."

As for the idea of no "absolute rest," this postulate seems to contradict common experience. For example, are you not at rest somewhere reading this book? We discuss the issue of no absolute rest in more detail. In Chapter Nine, we consider arbitrary choice of reference frame velocity. At rest, however, we can see from the postulate above that local measurement of timekeeping frequency will result in the same measurement whether we use mechanical, wind-up clocks, electric clocks, or atomic clocks. We also soon see that consistent time measurement is true for everyone in the universe. Timekeeping frequency unquestionably is a scalar that describes time.

4.5 Timekeeping Frequency is Scalar Time in Wave Motion Mechanics

In 1924, Louis de Broglie extended the idea in Einstein's photoelectric effect, where light waves are also given particle properties, to conclude that all matter has wave properties. Equation 4.4 is the basic formula for the speed of a wave:

$$V = f\lambda \qquad \text{Equation 4.4}$$

Equation 4.4 states that the speed of a wave, v, equals the wave source frequency f multiplied times the wave's wavelength λ. So frequency as a concept for scalar time is reasonable. Scalar frequency is not the idea for Universal Time T that we seek for measurement of motion. But, frequency demonstrates utility for the motion of waves. Frequency as scalar time illustrates the wave-particle duality of nature.

4.6 Frequency as Scalar Time Is Not Universal Time

This search for Universal Time timekeeping has been fruitful, but not successful. We now have a better idea of the need for Universal Time T to be a true scalar. Unfortunately, traditional time t has questionable qualities of past, present and future. Universal Time must be a scalar. It must be measurable to all who measure motion in each of four dimensions of physical space, x, y, z, and τ.

To better understand space we want to redefine time. Yet we need to recognize an important fact. The problems in Chapter One result from Einstein's idea of relative motion. Einstein states that all motion is relative to the observer. Einstein's theory selects the speed of the observer to be zero. This choice of zero speed of an observer appears to violate the postulate of special relativity, "...to the concept of absolute rest there corresponds no properties of the phenomena." Measurement of relative velocity in three-dimensional space creates time dilation and length contraction. Other complications add to the problem (Miller 1977 1981, 370).

A similar complication results when an observer at rest receives a sound, or an electromagnetic, communication of timekeeping frequency. Due to the Doppler shift, an observer will receive this frequency at a greater or lesser rate for respectively approaching or receding relative motion. Einstein shows how the Doppler shift can be derived from relative motion in his 1905 paper on Special Relativity.

So it appears that measurement of scalar time, as frequency of timekeeping, is locally limited to the environment of the timekeeper. For Einstein's relative motion, the observer's clock's timekeeping frequency has limits. Timekeeping frequency is not the scalar idea for Universal Time T we seek. Timekeeping frequency varies with the relative motion of the timekeeping clock.

Chapter 5
Universal Time
In Four-Dimensional Physical Space

"Imagination is more important than knowledge.
Knowledge is limited; imagination encircles the world."

--- Albert Einstein (Calaprice 1996, 228)

5.1 Frequency Is Scalar Time, But Not Universal Time

We have just seen from the previous chapter that timekeeping frequency as scalar time is not the concept needed for Universal Time T. The notion of timekeeping frequency as scalar time is too simple to meet our needs. Many reasons justify our conclusion:

◆ Frequency as scalar time measures in units of cycles per second, not seconds.

◆ The best use of frequency is a scalar time to measure waves' motion. See Equation 4.3, $v = f \lambda$.

◆ Frequency changes with relative motion. Moving clocks exhibit a Doppler shift in timekeeping frequency.

◆ Motion along each of four physical dimensions requires a Universal Time T. T is similar to Newtonian absolute time t. Time t measures motion in three dimensions of physical space.

◆ We need a Universal Time that exists in all parts of the universe as a synchronous and isochronic measure of scalar time.

Eddington's quote in Chapter One is again revealing. He observed that the modern principle of relativity comes from Einstein's insight. This insight is that local time was *the time* for the moving observer. This insight exceeded previous thought that the slowing of moving clocks is an illusion. Einstein's insight therefore is that no real, or absolute, time exists. The truth of whether absolute time exists depends upon the definition of time. If what clocks measure defines time, then time does not exist as Newtonian absolute time. And, measurement of time is relative. Dissecting time into Timespace τ and

Universal Time T make Newtonian-like absolute measurements of time and motion possible again. Now these Newtonian-like measurements and ideas apply to four-dimensional physical space. We find the solution to the paradox of timekeeping for a moving clock. The solution is simply the variation in the clock's motion through Timespace τ, the fourth physical dimension.

5.2 Fly in the Automobile and Universal Reference Frames

Until you ask someone to define time, everyone knows what time is. A Universal Time T that exists for each of four physical dimensions is even more elusive.

We start by considering a fly buzzing around inside a moving automobile in Figure 5.1:

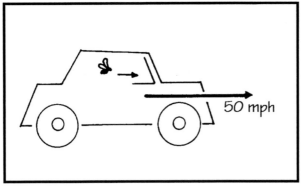

Figure 5.1 (G043)
Fly Inside an Automobile

This pesky fly is in your world of experience. The auto with you and the fly moves at 50 miles per hour. How can this small fly keep up with the speed of the auto? The current physics explanation uses Newton's first law of motion, the law of inertia. The law of inertia states that an object in motion continues in motion unless an external force acts on the object. A force must act on the fly it to slow up or speed up its motion. The fly will feel the force of any acceleration or deceleration. Yet, the fly is unaware of any motion of the automobile.

This explanation is not the most interesting part of the fly's motion. **What's interesting is that the fly's condition of a constant speed of 50 mph is equivalent to the fly at rest with no motion.** When you are in a room on Earth, you travel toward the room's east wall at a speed of approximately 1000 mph as the Earth rotates. You are glad that the east wall is also moving with this same speed. If the Earth would suddenly stop rotation, you would collide with the east wall at 1000 mph! The important question is whether your speed is zero or 1000 mph. Is your speed just an arbitrary choice? Conventional thought answers this by stating that your speed simply depends on your reference frame. This answer neglects that both reference frame conditions are identical. The answer assumes that space is three-dimensional.

Those in a **Universal Reference Frame observe no condition of change in the reference frame's motion.** The fly in the automobile and the person on Earth, experience a Universal Reference Frame. In a Universal Reference Frame, one might consider arbitrarily choosing the motion of the Universal Reference Frame. The explanation why a Universal Reference Frame's speed must be zero, as in Chapter Nine. Chapter Ten discusses Einstein's principle of equivalence. Einstein's equivalence principle applies to a Universal Reference Frame. A Universal Reference Frame's speed must be zero, even when the reference frame is accelerating with increasing speed.

Timekeeping in all Universal Reference Frames of the universe is synchronous and isochronic. Universal Time is consistent even if relative motion exists between any two Universal Reference Frames. Our insight here is just the flip side of the relativity paradox. As we learned from Eddington in Problem Nine of Chapter One, Einstein's insight was that observed time on clocks with relative motion was the "local time." (Eddington 1920 1959, 211) **Unfortunately from this insight of Einstein, a multitude of paradoxes emerge. In contrast, everyone in the universe identically measures Universal Time T in four-dimensional physical space!**

We redefine traditional time t. Time t now becomes two ideas, Timespace τ and Universal Time T. We confidently conclude that all observers in Universal Reference Frames measure Universal Time identically.

In 2004 Brian Greene, in *The Fabric of the Cosmos*, describes a passage of time that he calls "universal." He says that this absolute time refers to the age of the universe (Greene 2004, 235):

> "And, this means that, with respect to space itself, all the clocks are actually stationary, so they tick off time identically."

When we measure Universal Time T in Universal Reference Frames, we realize that these clocks need not be stationary, or at absolute rest. Instead the clocks can move relative to each other with constant speed or acceleration. The condition of a Universal Reference Frame's zero motion and the condition of a reference frame's constant speed motion is equivalent, or universal.

Einstein's principle of equivalence states that the condition of accelerated motion is equivalent to zero motion. Likewise with Universal Reference Frames, the condition of a reference frame's accelerated motion and the condition of a reference frame's constant speed motion are equivalent, or universal. **Einstein's principle of equivalence from general relativity becomes a Universal Principle of Equivalence**.

Einstein's postulate of special relativity says there is no property of absolute rest. If there exists no property of absolute rest, then absolute motion cannot exist. The nonexistence of absolute motion follows if absolute motion is relative to nonexistent absolute rest. Einstein's non-Euclidean geometry of time and space concludes that motion is not absolute, but is relative to the observer. Einstein concludes that absolute motion doesn't exist.

In contrast, Universal Reference Frames are at rest. Motion in these reference frames is absolute motion. Symbols v_s and v_r describe this absolute motion in equations.

5.3 Universal Time T is a Scalar

We can measure Universal Time T as scalar time in a Universal Reference Frame. How can a clock's measure of time be a scalar measure of time? In Chapter Four, we saw how frequency as scalar time met the necessary condition of no direction in physical space. Measuring Universal Time in seconds, hours and years also meets this condition.

The relationship between traditional Newtonian absolute time t and frequency f appears in Equation 5.1:

$$t = \frac{1}{f} \qquad \textbf{Equation 5.1}$$

Here f is the frequency of vibration in cycles per second, and t is the time in seconds it takes for one cycle. This t is the "period" of vibration. More properly, t's unit of measure is seconds per cycle (s/cycle). The unit of measure for frequency f is cycles per second (cycles/s).

So the traditional measure of time is just a multiple of periods recorded on a clock. Time is a count of timekeeping periods. Considering time as a multiple of timekeeping periods makes time a scalar measurement. In physical space time has no quality of direction. If timekeeping frequency is a scalar, then time is a scalar in traditional Newtonian physics. As stated earlier in Chapter One, time as a scalar is contrary to the thinking of introductory students, who incorrectly consider time having past and future directions. In a similar way, Universal Time T is a scalar quantity.

5.4 The Nature of Universal Reference Frames

The issue of absolute motion and of the arbitrary choice of a reference frame's speed is resolved in Chapter Nine. The discussion there enlightens you with a unique solution. Chapter Nine also shows that Euclidean geometry, using Timespace and Universal Time, allows absolute motion to exist in Universal Reference Frames. These Universal Reference Frames are at rest with zero motion.

Universal Reference Frame qualities need more clarity. In his words, Newton states the first law of motion as the law of inertia (Hutchins 1934 1952, 5):

"Every body continues in its state of rest, or of uniform motion in a right [straight] line, unless it is compelled to change that state by forces impressed upon it."

An Inertial Reference Frame is any reference frame in which Newton's first law is valid (Giancoli 1984 2000, 917). This definition limits an inertial reference frame to straight line motion with zero, or constant, speed. This definition does not include the case of accelerated motion. Nor does Newton's first law include a centripetal force and acceleration, maintaining constant speed in circular motion, where the velocity's direction changes at a constant rate.

A Universal Reference Frame is any reference frame that locally identifies zero reference frame speed. The following are different examples of a Universal Reference Frame's motion:

- ◆ **Zero motion.**
- ◆ **Straight-line motion with constant speed.**
- ◆ **Circular motion with constant acceleration and constant speed.**
- ◆ **Accelerated motion of all types.**

A Universal Reference Frame is universal in every way. Any clock's scalar time within any Universal Reference Frame defines **Universal Time T.** Description of absolute motion within a Universal Reference Frame results by locally identifying the reference frame's speed to be zero.

It is difficult to accept the idea that the total speed in four dimensions is the speed of light. Equation 3.1 and Figure 3.1 symbolize this idea. There is another way to understand the significance of Universal Reference Frames. Consider that three-dimensional xyz S-space moves at the speed of light in the direction of Timespace. **Just as the fly in an automobile cannot perceive the motion of the auto, we in xyz S-space cannot perceive our motion along the dimension of time, called Timespace.** Only forces with acceleration and deceleration can be perceived and measured, not motion. For example, the force we perceive toward objects of mass, Newton calls gravity and Einstein calls warped space.

To identify Universal Reference Frames and the measurement of Universal Time T, we consider the following four examples:

◆ Timekeeping in an auto moving with constant speed.
◆ Timekeeping in a moving spaceship very far away from any massive planet.
◆ Timekeeping in a laboratory room on the surface of the Earth.
◆ Timekeeping on a wristwatch that is in a free-falling, broken elevator.

Timekeeping in an auto moving with constant speed: The fly inside an auto leads us to conclude that the state of zero motion and the state of constant speed motion are the same. Timekeeping in a moving Universal Reference Frame is identical to timekeeping in a stationary roadside Universal Reference Frame. Both Universal Reference Frames have zero speed. These clocks measure Universal Time T. Note in this simple example, we did not consider any vertical forces or accelerations due to gravitation.

Timekeeping in a moving spaceship very far away from any massive planet: Timekeeping in a moving spaceship very far away from any massive planet creates weightlessness. Regardless of whether the speed of this spaceship is constant, or accelerating with the force of the spaceship's engines according to the Universal Principle of Equivalence, the timekeeping on this spaceship would be in a Universal Reference Frame. Timekeeping by the weightless astronaut would measure Universal Time T.

Timekeeping in a laboratory room on the surface of the Earth: On the Earth, timekeeping is timekeeping within a Universal Reference Frame. Again, a speed of zero, or a speed of 1000 mph toward the room's east wall, are equivalent conditions of a Universal Reference Frame. From an external view away from Earth, Earth's rotational motion produces a surface speed of 1000 mph. A centripetal force of gravity pulling down creates this motion. A small part of this downward gravitation is the centripetal force creating the circular motion. The upward force on objects balances the remaining downward pull of gravity we call the weight of the object. Vertically the forces balance, with only a small component of gravity pulling down creating constant 1000 mph circular speed. This small centripetal acceleration pulling toward the center of the Earth produces accelerated circular motion of the laboratory reference frame. According to the Universal Principle of Equivalence this acceleration of the reference frame creates a Universal Reference Frame. Laboratory room clocks are in a Universal Reference Frame and they measure Universal Time T. Laboratory room clocks locally identify zero reference frame speed.

Timekeeping on a wristwatch that is in a free-falling, broken elevator: Timekeeping in a free-falling, broken elevator is Universal Time T. Einstein's principle of equivalence from general relativity theory says that a free-falling person cannot distinguish the external force of gravitation. The descending person perceives this free-falling local condition as a speed of zero. According to Einstein, as the person falls with the elevator, or moves with accelerated motion, the observed rate of timekeeping by an external observer varies with gravitational acceleration increasing the speed of fall. With new definitions for time, Timespace and Universal Time, we understand that this free-falling person locally identifies zero reference frame speed and measures Universal Time T on a wristwatch. **Universal Time T is a new way, a universal way, to use Einstein's principle of equivalence**. This free-falling person gathers speed as the person falls. This free-falling person is in an accelerating Universal Reference Frame. Yet, Universal Reference Frames are universal. Universal Reference Frames include reference frames that accelerate.

In the general theory of relativity, as an alternative to Newtonian universal gravity, Einstein introduces the idea of warped, three-dimensional space. Einstein describes the free-falling person as experiencing warped, three-dimensional space rather than a Newtonian force of gravity. Einstein's theory recognizes that the rate of timekeeping varies with radial distance away from an object with mass, like a planet. The rate of timekeeping, as speed through Timespace v_τ and as with Einstein's general relativity theory, remains uniform with clocks at a given distance or orbit from a central mass M. A general equation for this condition is in Chapter Ten.

5.5 Universal Timekeeping Frequency F and Timekeeping Rate v_τ

With the creation of the new ideas of Timespace τ and Universal Time T frequency measurement must also change. A new concept of Universal Timekeeping Frequency F necessarily follows. Universal Time T establishes the existence of consistent Universal Timekeeping Frequency F. Universal Timekeeping Frequency F is fundamental to the new ideas for time we present here.

The reciprocal of the timekeeping period, or the traditional time t of timekeeping vibration, is the definition of traditional frequency, f. We see this relationship in Equation 5.1. In a similar way, **Universal Timekeeping Frequency F is the reciprocal of the period of timekeeping with Universal Time T.** Equation 5.2 illustrates the nature of Universal Timekeeping Frequency F and timekeeping rate v_τ:

$$F = \frac{1}{T} = \frac{v\tau}{\tau} = \frac{c}{\tau_O} \qquad \text{Equation 5.2}$$

As expected, Equation 5.2 shows that Universal Timekeeping Frequency F is the same everywhere in the universe. When the speed of an object is zero or $v_s = 0$, then the object's displacement through Timespace is τ_0 and the object's speed through Timespace is at maximum speed c, the speed of light. This condition is the last term in Equation 5.2. Equation 3.1 describes that an object's speed of v_s in xyz S-space corresponds to a speed of v_τ through Timespace τ.

Einstein's relative motion produces a variation in observed timekeeping frequency f, known as the Doppler shift. Universal Timekeeping Frequency F is the same everywhere in the universe. **Our new ideas for time lead us to v_τ as a measure for the variation in timekeeping rates. The speed v_τ through Timespace is the rate of timekeeping. This variable was not available to Einstein when he developed his equation for time dilation. So, Einstein limited the idea of variable rates of timekeeping to his equation for time dilation. See Equation 1.1.**

As for the Doppler shift in traditional physics, a variation in received frequency results from the relative motion of the wave source. Although sound waves and light waves are different, both produce a variation in wave frequency with relative motion of the wave source . For the new ideas here, there is no variation in Universal Timekeeping Frequency F for objects in relative motion.

For objects in relative motion, instead of finding a variation in timekeeping frequency, we find a variation in the rate of speed v_τ through Timespace τ. Doppler shift equations in traditional physics determine the variation in observed frequency as a function of the wave source's speed. Our new ideas produce an analogy to the Doppler shift equation. Rearrange the four-dimensional motion Equation 3.1. Equation 3.1 states the total speed of all objects in physical space is c, the speed of light. Equation 5.3 describes the variation in the rate of timekeeping observed for an object moving at speed v_s:

$$V_\tau = \sqrt{c^2 - V_s^{\,2}} \qquad \textbf{Equation 5.3}$$

Chapters Eight and Ten respectively present Euclidean timekeeping rate equations for black holes and planets. These equations produce three different solutions:

- These equations describe the varying rate of timekeeping v_τ for clocks near an object with mass.
- These equations describe absolute motion traditionally used in physics.
- These equations eliminate Einstein's relative motion paradoxes.

5.6 Clocks Measure Timespace Displacement in Meters

Einstein's relativity theory concludes that clocks measure time at different rates. How can clocks measure a Universal Time T that is consistently the same everywhere? If you think a clock's timekeeping is slower for clocks in relative motion, you neglect the universal quality of Universal Reference Frames. You limit the definition for time to what a clock measures.

Clocks in Universal Reference Frames measure more than Universal Time in seconds and hours. These same clocks also measure Timespace displacement in meters. These physically moving clocks record variations in Timespace displacement in meters due to the clocks' varying timekeeping rates, or speed through Timespace.

Instruments can measure more than one kind of physical measurement. For example, the barometer measures atmospheric pressure in millimeters of mercury. The barometer also measures the altitude of airplanes in flight in meters above the ground. The altimeter is the new name for this barometer use. Atmospheric pressure uniformly decreases with increasing height above the ground. This makes the altimeter work.

In an analogous way, a moving clock measures a decreasing rate of displacement through Timespace. The displacement through Timespace decreases, as the speed through xyz S-space increases. Instead of minutes, a clock face can show a system with kilometer units. The speed of the clock through Timespace determines the system of a clock face's kilometer units.

A clock face measuring Universal Time T has minute units. Measuring Universal Time T locally applies to clocks in Universal Reference Frames.

Chapter 6
Lorentz-FitzGerald Contraction of Timespace
Is Time Dilation

"People do not realize how great was the influence
of Lorentz on the development of physics."

--- Albert Einstein (Calaprice 1996, 72)

6.1 The Lorentz-FitzGerald Contraction Implies Constant Light Speed

George Francis FitzGerald explained the failure of the famous, 1887 Michelson-Morley experiment (Asimov 1964 1972, 464). This experiment showed for any motion of a source of light, the speed of the light is always the same value. Motion of a light source can't increase the light's speed.

To understand this idea, first consider a fly outside of an automobile. As the auto moves at 50 mph, the fly approaches the auto's windshield at one mph:

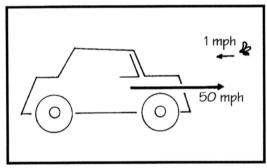

Figure 6.1 (G027)
Fly and Automobile Collide

The fly and the auto's driver have very different experiences as the fly splats against the windshield. Yet, each measures the same relative speed of the oncoming object to be 51 mph.

Now consider Spaceship A and Spaceship B, approaching each other and firing laser light beams. See Figure 6.2.

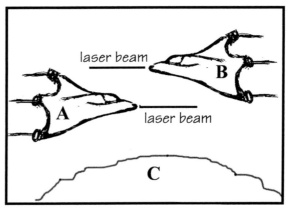

Figure 6.2 (G041)
Spaceship A and Spaceship B

A pilot in Spaceship A measures the approaching laser beam to have a speed of c = 186,000 miles per second. The pilot in Spaceship B measures the same value for the laser beam approaching from Spaceship A. Even though these two spaceships move at vast speeds. Even with spaceship speeds near the speed of light, the result will be the same. Likewise, an observer on Planet C measures each laser beam from these spaceships to have this same speed. With the measurement of the speed of light constant, space and time are no longer as simple as the relative motion of the auto and the fly.

FitzGerald concluded in 1895 that measurements of the length of moving objects, and the distances they travel, must contract as they move. He reached this conclusion to theoretically maintain a constant measure of light speed. His length contraction formula made constant light speed possible without considering Einstein's 1905 time dilation. On the next page Equation 6.1 shows the traditional physics formula for contraction:

$$l = l_o \sqrt{1 - \frac{v^2}{c^2}}$$

Equation 6.1

The symbol "l" is the contracted length of the moving object. The symbol "l_o" is the measured length of the object at rest. If an object moves through xyz S-space at speed v less than c, the speed of light, the square root in Equation 6.1 becomes a fraction smaller than one. The measurement of length of anything moving at speed v contracts to a size smaller than its length l_o, measured at rest.

A person in motion measures the speed of light with the same result as a person at rest would measure. A person in motion with the moving object does not measure contracted object length because this person and the object are not moving relative to each other. Different time measurement for these two observers is a paradox that Einstein's theory requires.

Hendrik Antoon Lorentz also tackled the experimental results of the Michelson-Morley experiment (Asimov 1964 1972, 477). Lorentz's theoretical work with electricity and magnetism also concluded that a contraction must take place. Lorentz based his conclusion on electron motion. He further declared that the maximum speed of an electron is the speed of light. In 1905, Einstein declared that the maximum speed of any object is the speed of light.

Einstein's 1905 special theory of relativity was the first theory to examine the consequences for time measurement when objects move. Einstein showed that timekeeping slows down for moving objects. Chapter One describes his time dilation in more detail. Time dilation was the genius of Einstein.

6.2 Derivation of the Formula for the Contraction of Timespace

Does light move through time? The answer to this question is less clear than answering the question: Does light move through Timespace? Timespace has similar qualities to dimensions of xyz S-space. Light travels through Timespace just as it travels through xyz S-space. If we give the fourth dimension properties of physical space, we answer: Yes, light moves through Timespace. Timespace is not visible. But, neither is time.

Each xyz S-space dimension has similar qualities to the fourth dimension of Timespace. Timespace must contract just as the three dimensions of x, y and z contract. Equation 6.2 describes Timespace contraction. Timespace contraction depends on speed in xyz S-space like contractions of xyz S-space. The Timespace contraction formula in Equation 6.2 does not depend on

Timespace speed v_r. Contractions of dimensions greater than three dimensions depend on the speed in xyz S-space. Motion through Timespace, timekeeping rate v_r, depends on motion through xyz S-space in Equation 3.1. Motion through xyz S-space determines motion in Timespace τ. Equation 6.2 shows that the contraction of Timespace is similar to the contraction of xyz S-space:

$$\tau = \tau_o \sqrt{1 - \frac{v_s^2}{c^2}} \qquad \text{Equation 6.2}$$

6.3 Contraction in Each of the Four Dimensions of Physical Space

Timespace and Universal Time use four Euclidean dimensions of physical space. Equation 6.2 through Equation 6.5 describe contractions of each one of these four dimensions of physical space:

$$x = x_o \sqrt{1 - \frac{v_x^2}{c^2}} \qquad y = y_o \sqrt{1 - \frac{v_y^2}{c^2}}$$

$$\text{Equation 6.3} \qquad\qquad \text{Equation 6.4}$$

$$z = z_o \sqrt{1 - \frac{v_z^2}{c^2}} \qquad \tau = \tau_o \sqrt{1 - \frac{v_s^2}{c^2}}$$

$$\text{Equation 6.5} \qquad\qquad \text{Equation 6.2}$$

In these four equations x, y, z and τ are contracted measures of moving objects' dimensions or displacements. The variables of x_o, y_o, z_o and τ_o are measures corresponding to objects' dimensions or displacements while at rest in xyz S-space. Appropriate displacement units, such as meters or miles describe each measurement. The new idea of Timespace τ is four-dimensional displacement due to the motion of time's passage, commonly called aging.

6.4 Method One: Timespace Contraction is Equivalent to Time Dilation

The Lorentz-FitzGerald contraction of Timespace, the fourth dimension, is algebraically equivalent to the idea of Einstein's time dilation. The total speed of all objects in four-dimensional space is the speed of light. Equation 6.6 shows two equivalent expressions for Timespace τ:

$$\tau = \tau_o \sqrt{1 - v_s^2/c^2} = v_\tau T \qquad \text{Equation 6.6}$$

The first expression for Timespace τ in Equation 6.6 represents the Lorentz-FitzGerald contraction of the fourth dimension. The second expression for Timespace τ in Equation 6.6 equals the speed through Timespace multiplied times Universal Time T.

Equation 6.7 is an expression for Timespace in a Universal Reference Frame. In a Universal Reference Frame, the reference frame's velocity is zero in xyz S-space and the maximum speed c, the speed of light, in Timespace:

$$\tau_o = cT \qquad \text{Equation 6.7}$$

Combining Equations 6.6 and 6.7, Universal Time, T, disappears. Equation 3.1 from Chapter Three results. So in four dimensions, all objects move with the speed of light at all times:

$$v_s^2 + v_\tau^2 = c^2 \qquad \text{Equation 3.1}$$

6.5 Method Two: Timespace Contraction is Equivalent to Time Dilation
Einstein's time dilation of Universal Time T gets an algebraically identical result. Equation 6.8 is Einstein's time dilation equation for Universal Time T. Equation 6.9 represent the condition of zero speed in xyz S-space. Combine the two equations. T_o disappears. Again we get Equation 3.1:

$$T = \frac{T_o}{\sqrt{1 - v_s^2/c^2}}$$

Equation 6.8

$$\tau = v_\tau T = cT_o$$

Equation 6.9

$$v_s^2 + v_\tau^2 = c^2$$

Equation 3.1

6.6 Method Three: Timespace Contraction is Equivalent to Time Dilation
A simple algebraic rearrangement of Equation 3.1 reveals that the Lorentz-FitzGerald contraction coefficient is equal to v_τ/c:

$$v_s^2 + v_\tau^2 = c^2 \qquad \text{Equation 3.1}$$

$$\sqrt{1 - v_s^2/c^2} = v_\tau/c \qquad \text{Equation 6.10}$$

Rearrange Equation 3.1 to get Equation 6.10. With increasing motion in xyz S-space, motion in Timespace decreases to a fractional value of its maximum speed c, the speed of light. So, contraction of Timespace equals the fractional decrease, v_τ/c, in the rate of the passage of time. The speed of light equals its displacement divided by the time of travel. **So, the fractional decrease in the measurement of time combines with the same fractional decrease in displacement in xyz S-space. Regardless of reference frame motion, the speed of light will always have the same value.**

Position 1 through position 5 in Figure 6.3 illustrates an increase in xyz S-space speed from zero to the speed of light at position 5. Figure 6.3 also illustrates the corresponding decrease in the speed through timespace from the speed of light, c, at position 1 to zero speed at position 5.

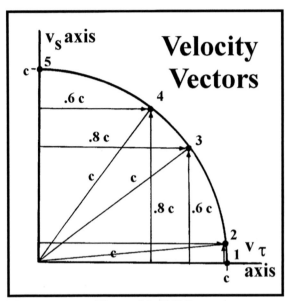

Figure 6.3 (G039)
Velocity in Space vs. Velocity in Timespace

6.7 Contraction of the Fourth Dimension Replaces Need for Time Dilation
Now we've introduced three salient notions:
- ◆ The fourth dimension of Timespace contracts with motion in three dimensions of xyz S-space.
- ◆ All objects move at the speed of light at all times.
- ◆ Timespace and Universal Time replace Einstein's time dilation.

New ideas of Timespace and Universal Time eliminate the necessity for the confusing paradoxes from Einstein's relative motion and its time dilation. Timespace describes the passage of time and measures time displacement in meters. Contraction of Timespace τ with xyz S-space motion is equivalent to Einstein's slowing of the traveling clock in relative motion. Universal Time is similar to the absolute time used by Newton. In Universal Reference Frames everywhere, timekeeping measures the same Universal Time. Surprisingly, a Euclidean description of four-dimensional space is simple and reasonable.

Chapter 7
What Does Four-Dimensional Space Look Like?

Esau
"I saw Esau kissing Kate,
The fact is we all three saw;
For I saw him,
And he saw me,
And she saw I saw Esau.

--- *I Saw Esau: The Schoolchild's Pocket Book*
Edited by Iona Opie, Walker Books, 1992
(Permission by Iona Opie)

Four-Dimensional Version of "Esau"
"I saw He saw Kissing Kate,
The fact is we all three saw
For I saw him,
And he saw me,
And she saw I saw He saw."

These two versions of the same children's rhyme are a fun analogy of how four-dimensional physical space hides within our experience with three-dimensional physical space. Esau, Kate and I are the three persons in the rhyme, *Esau*. In the four-dimensional version of *Esau*, three persons saw Kissing Kate. This makes a total of four persons. But, the use of "he" and "him" in the four-dimensional version of the rhyme does not identify who looked and who saw. Visually seeing four-dimensional space is just as elusive.

7.1 What Does Four-dimensional Space Look Like?
Every reader who reads descriptions of four-dimensional space silently asks the question: "What does four-dimensional physical space look like?"

Over the last two centuries, many good books attempt to describe four dimensions and what this space must be like. Authors include Abbott and Burger (Abbott and Burger 1880 1994), Ellis (Ellis 1988), Hinton (Hinton

1904)(Hinton 1980 1993), Rucker (Rucker 1976 1977) (Rucker 1880 1980) (Rucker 1984) (Rucker 1985), and Stewart (Stewart 2001).

Hilbert unfortunately uses a mathematical error in his attempt to describe four-dimensional space. He considers angular measurement, ds, equivalent to distance measures, du and dv. He uses the calculus equation, $ds^2 = du^2 + dv^2$, to conclude that four dimensions are in a torus shape, or donut (Hilbert and Cohn-Vossen 1952 1999, 341). The correct expression is the trigonometric expression: $\tan s = u/v$. The desire to visually describe four-dimensional space is strong. Although authors take many approaches, no agreement on a single, clear and definitive vision of four-dimensional space results.

I challenged students of science and members of public audiences to consider the new ideas presented here. I asked them to consider a practical knowledge of four-dimensional space. Eleven years of thought created this book of new ideas.

7.2 Euclid's Geometry and Binocular Vision

First consider what xyz S-space looks like. Instead of thinking about it from your perspective, consider what space looks like to three different individuals:

- ◆ A non-seeing person without sight from birth.
- ◆ A person with sight in only one eye.
- ◆ A person with the complete binocular vision of two eyes.

Ask the non-seeing person what space looks like: "Sighted people can just look at objects and know how far away they are. How is this possible?" Answers to this question vary: "I do not know," and "Perhaps objects are more blurry at a distance." The non-seeing person cannot formulate a mental concept for what we mean by the term, space, because this person does not have the benefit of sight.

The person with sight in only one eye sees two-dimensional space. This person often has trouble driving a car. With sight in only one eye, a person lacks depth perception. This person cannot see space.

The person with binocular vision has a brain that uses two different images, one from each eye. In this way, the person perceives depth. Three-dimensional movies use paper glasses with two different colored lenses. These lenses cause the moviegoer to see a different image with each eye. The brain then fabricates a three-dimensional monster, stepping out of the movie screen and into the room of the theater.

Gedankin Experiment: So what do we need to see the fourth dimension of Timespace? A third eye? Special four-dimensional glasses? A third eye is not a requirement for seeing four dimensions. After careful consideration, only one clear answer exists. Beyond two eyes, we also need new ideas created by our minds to see four-dimensional physical space.

Figure 7.1
Three Eyes Cannot See Four
Dimensions

Philosopher S. Alexander talks about the role of the mind in the creation of four-dimensional space (Alexander 1920 1950, 160):

"The construction of a more than three-dimensional space does not lead us into a neutral world but takes notions which are empirical at bottom and combines them by an act of our minds."

The description of four-dimensional space in Einstein's non-Euclidean geometry is not visual. Non-Euclidean geometry uses algebra and higher mathematics. The new ideas of Timespace and Universal Time are only an attempt to visualize four-dimensional space using Euclidean geometry.

Euclid wrote two classic works, one about three-dimensional geometry and one about vision. "Seeing is believing." Euclid's geometry can provide us with a limited, but satisfying approach, to understanding four-dimensional physical space that openly rejects vision for enlightenment.

7.3 The Passage of Time is Not Visual

Just as you cannot see time, you cannot see four dimensions. "What does Timespace's fourth dimension look like?" That is like asking: "Can you see the passage of time?" If you look at your watch, you can see the motion of the second hand as it sweeps around the dial. Motion is all you can see.

Timespace in meters, is a mental construct that represents the passage of time. Timespace offers many qualities that parallel the three dimensions of x, y, and z in S-space. Age, or Timespace displacement, is a vector concept of the passage of time, not a scalar concept of time. If you are 39 years old, and haven't traveled much, your speed in the direction of Timespace is always the maximum, the speed of light = 186,000 miles/second. Measured in Timespace you are 230 trillion-miles old:

$$(39 \text{ years}) (31,536,000 \text{ seconds/year})(186,000 \text{ miles/second}) =$$
$$230,000,000,000,000 = 230 \text{ trillion-miles old!}$$

Not very convenient? Using Timespace displacement as an age measurement would be like moving to a country where inflation is out of control. There, buying one loaf of bread requires a large bag of paper money.

Four-dimensional Euclidean geometry with Timespace as the fourth-dimension provides a new opportunity. Later in this chapter we see the Euclidean advantage that allows us to see four-dimensional space.

All traditional attempts fail to visualize four dimensions. Authors use four-dimensional, non-Euclidean geometry in Einstein's general theory of relativity and describe the universe as "warped." They use models of lesser dimensions like the two-dimensional picture of warped space in Figure 7.2:

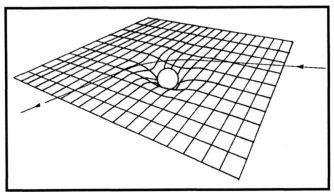

Figure 7.2 (G060)
Two-Dimensional Warped Space

Of course, binocular vision of a warped four-dimensional space is not possible. Using two-dimensional analogies, such as Figure 7.2 and the Minkowski light cone in Chapter One, leads students of relativity theory to confusion and even frustration.

7.4 Lower-Dimensional Projections Cannot Adequately Describe Higher-Dimensional Space

Here's a much clearer model to help us visualize four-dimensional space. Lower dimensions are just projections from higher dimensions. For example, Figure 7.3 shows three different two-dimensional objects projecting onto one-dimensional space axes. Figure 7.4 shows one three-dimensional object, a pyramid projecting onto three different two-dimensional planes.

Projections from higher space in both illustrations provide incomplete knowledge of the higher dimensional objects. In fact, if you only have knowledge of lower dimension projections, then you have limited knowledge of higher dimension shapes. For example, the three-dimensional object in Figure 7.4 shows two projected triangles and one projected square. In a similar way, three-dimensional projections from four-dimensional space provide a limited knowledge of this non-visual, higher space.

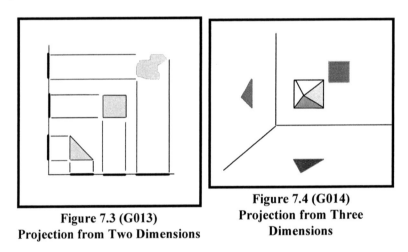

Figure 7.3 (G013)
Projection from Two Dimensions

Figure 7.4 (G014)
Projection from Three Dimensions

Everyone has trouble understanding three-dimensional warped space in Einstein's general theory of relativity. Warped three-dimensional space is also a non-visual abstraction. Warped three-dimensional space is an inadequate attempt to understand the geometry of four-dimensions.

Prose by Michio Kaku in *Beyond Einstein* vividly describes that only three-dimensional projections can be visible from higher dimensional space (Kaku 1987 1995, 169):

"As spheres of flesh grabbed us and flung us into higher-dimensional space, we would see only three dimensional cross-sections of the higher universe."

The four-dimensional hypercube in Figure 7.5 is strange looking. Yet, the hypercube meets mathematical requirements for a four-dimensional object.

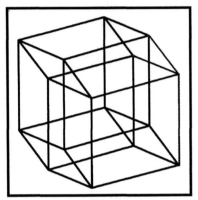

Figure 7.5 (G015)
Three-Dimensional Hypercube

Yet, the hypercube is an object within three-dimensional space. When a three-dimensional hypercube projects an image onto a flat screen. The two-dimensional image on the flat screen is that of a three-dimensional cube! See Figure 7.6. A hypercube projects a three-dimensional image onto a two-dimensional screen:

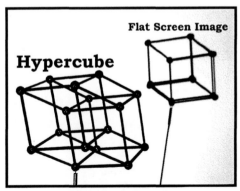

Figure 7.6 (G087f)
Photograph of Zoomtool Hypercube
Projection on a Flat Screen

The hypercube allows us to understand that a visible four-dimensional object can be created in three-dimensional space. But, vision of four-dimensional space remains unobtainable.

7.5 Four Three-dimensional Projections from Four-dimensional Space

Minkowski first showed the four-dimensional nature of space created by Einstein's special theory of relativity. Minkowski called this four-dimensional space a space-time continuum. His space-time continuum used non-Euclidean geometry. Later in Chapter Nine an explanation of non-Euclidean geometry appears.

With four-dimensional physical space of x, y, z, and Timespace τ, four different three-dimensional projections from higher, four-dimensional space result. Timespace and Universal Time use simple Euclidean geometry. This geometry describes four-dimensional space with four three-dimensional projections in Figures 7.6, 7.7, 7.8 and 7.9.

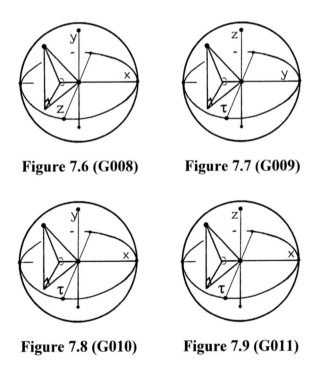

Figure 7.6 (G008)   **Figure 7.7 (G009)**

Figure 7.8 (G010)   **Figure 7.9 (G011)**

Gedankin Experiment: How many unique three-dimensional images project down from five-dimensional space? The answer is twenty different three-dimensional projections. Five-dimensional space produces five projections of four-dimensional space. Each four-dimensional space produces four three-dimensional projections. Answer: 5 x 4 = 20. This result is just like each three-dimensional space produces three two-dimensional projections as illustrated in Figure 7.4.

Superstring theory uses eleven dimensions of higher-dimensional space---10 dimensions of space with one dimension of time. How many three-dimensional images project down from Superstring's eleven-dimensional space? Answer: 11 x 10 x 9 x 8 x 7 x 6 x 5 x 4 = 6,652,800 three-dimensional projections! This huge, unmanageable result makes one see the mathematical complexities of non-Euclidean geometry as a preferred approach to superstring theory.

7.6 The Fantastic Twin Trip in Three Dimensions!

With Euclidean geometry we can clearly illustrate the fantastic twin trip. Consider a simple trip along the x-axis out and back for the traveling twin. During this motion the stationary Earth twin remains on Earth. Figure 7.11 on the next page illustrates what happens in one three-dimensional, x-y-τ projection. Twin E is the Earth-bound twin. Twin T is the traveling twin.

The projection in Figure 7.11 on the next page describes only x-space, y-space, and Timespace τ. In this trip, the traveling Twin T goes out and comes back along the x-axis. We will use the convenient values that when either v_s or v_τ equals .8c, then corresponding values for the other speed equals .6c. For simplicity both Figures 7.10 and 7.11 illustrate these convenient values for speed:

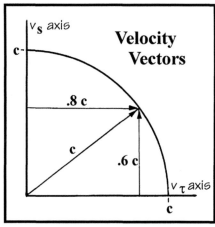

Figure 7.10 (G037A)

Figure 7.11 (G063f)

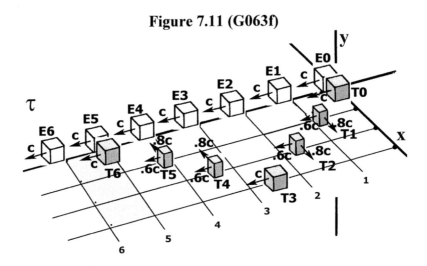

The twin trip illustrated in Figure 7.11 is a sequence of these events:
Twin T accelerates away from Earth along the x dimension, T0 to T1.

- ◆ Twin T moves away along the x axis at constant speed, T1 to T2
- ◆ Twin T decelerates to a stop, T2 to T3, and stops at T3
- ◆ Twin T accelerates back to Earth along the x-axis, T3 to T4
- ◆ Twin T moves at constant speed on the return trip, T4 to T5
- ◆ Twin T decelerates to a stop on Earth from T5 to T6
- ◆ Twin T stops at T6 to reunite with E6 at x=0
- ◆ At the x=0 reunion, E6 is now older than T6.

At position 6 both E and T are at the same position in xyz S-space, but are at
different positions in time!

Earth Twin E, always proceeds through Timespace at c, the speed of light. Earth Twin E's speed in xyz S-space is always zero. Traveling Twin T's motion along the x-axis results in decreased motion in Timespace. In other words, Twin T moves at less than the speed of light in the direction of Timespace. When twins unite at x equal zero, Twin T is age 5 (T6) and Twin E is age 6 (E6).

With traditional physics we understand this slowing down in the direction of Timespace as observing that the traveling Twin T ages less. **In this three-dimensional projection from four-dimensional space, motion along a fourth dimension, called Timespace, replaces the confusing space-time continuum and the twin paradox**.

Figure 7.10 uses a cube shape to represent each twin. This cube shape allows us to see the contractions of dimensions for this twin trip. Contraction occurs only in the dimensions of x and Timespace τ and only when the traveling Twin T is moving relative to Earthbound Twin E. The contraction of the cube in the dimension of Timespace τ is not a visual contraction, but a contraction represented with reduced speed in the direction of Timespace. Note that Traveling Twin T at T3 has stopped for the instant when it turns around for the return trip. More important, the measurement of Universal Time T by Twin T for positions T0 to T6 is identical for each position. Twin T experiences a force of acceleration and a force of deceleration. But, in each condition T0 to T6, Twin T cannot locally observe any motion, even when accelerating!

Einstein's special relativity and its twin trip paradox do not consider accelerations of the spaceship. Einstein's general relativity theory provides the needed solutions for accelerated motion. **It is important to understand that in our Euclidean geometry approach accelerated motion is an acceptable motion in a three-dimensional projection from four-dimensional space!**

Three remaining projections, (1) x-y-z (2) x-z-τ and (3) z-y-τ, also describe the trip of Traveling Twin T. That is, a Euclidean approach to four-dimensional space is the sum of four different, three-dimensional projections. Yet these four projections do not reveal all of the richness of four-dimensional space.

Visualization of four-dimensional space is not possible. We can't see four-dimensional space just as we can't see the passing of time. Nor is non-Euclidean geometry in Einstein's relativity visible. Nevertheless, authors continue to describe space as warping of three-dimensional space.

Chapter 8
Time Travel, Time Reversal, and Astronomy

"Let's do the time warp again."

--- Rocky Horror Picture Show

8.1 Time Travel Is a Popular Topic

Almost countless references discuss time travel and time reversal. The popular print and electronic media continually present the possibility of time travel, even the Wall Street Journal (Begley 2003). Interesting reading about time travel and time reversal is in the following references: (Boslough 1992, 179 and 208) (Coveney and Highfield 1990) (Davies 1995 1996, 233 and 249) (Davies 2002) (Fraser 1982, 179) (Gott 2001) (Kaku 1987 1995, 178) (Kaku 1997, 342) (Nahin 1993) (Morris 1985) (Price 1996, 170) (Reichenbach 1956 1999) (Rucker 1984, 133) (Savitt 1995 1998) (Shallis 1983, 63) (Sklar 1985, 305) (Sklar 1974 1976) (Tulka 1977, 177) (Whitrow 1980, 321).

8.2 Time Travel and Time Reversal are Different Ideas

We can illustrate time travel with examples of the twin trip story. Time travel means a newborn twin can traverse xyz S-space at vast speeds, aging slowly. Upon return to Earth, the traveling twin aged five years. The Earth twin aged 70 years. Over the last century, experiments agree with Einstein relativity that time travel is possible (Will 1986). These experiments include atomic clocks on jet planes and fast-moving, radioactively decaying isotopes.

Einstein knew that the validity of relativity theory relied on experimental verification. Einstein's own words best describe this fact:

"No amount of experimentation can ever prove me right;
a single experiment can prove me wrong."
--- Albert Einstein (Calaprice 1996, 224)

So far all experiments agree with the time travel that relativity theory predicts!

Time reversal is a different story. Time reversal occurs when the timekeeping clock moves backwards. No experiment yet validates time reversal. The following sections carefully illustrate the difference between time travel and time reversal.

8.3 Time Travel is Aging at Different Rates

In another fun example, consider a child who is five years old and has 25-year-old parents. The parents take a trip together on a vastly fast moving spaceship. When they return to Earth, the child is 90 years old, and the parents are only 30 years old. Is this paradoxical aging time reversal or time advancing into the future? Time travel requires passage of time.

As the parents exit the spaceship door, they travel forward in time to Earth and their child. The parents' experience is like jumping 60 years into the future. But the parents find they and everything that they took on the trip have aged only five years.

Now, suppose that the 90-year-old child moves into the grounded spaceship. The child's experience is like moving 60 years backwards in time. The 90-year-old child now needs geriatric care. The parents will now be caring for their aging child! After caring for the baby's first five years, this is not fair. This new twist to the twin trip paradox makes an important point. Is this really time travel, or just aging at different rates? The answer is aging at different rates. Since both child and parents aged at different rates, authors sometimes call this reunion an example of time travel.

Why is this fun idea of aging at different rates out of our worldly experience? Well, consider the twin's trip to be limited to our fastest mode of transportation. Place a newborn twin in a 747-jet moving at maximum speed, constantly being refueled in the air. While the Earth twin ages to 70 years old, **the traveling twin will be less than two minutes younger upon returning to meet the Earth twin**. So, aging at different rates requires moving at vast speeds, greater than available transportation technology, speeds near the speed of light. Time travel as a physical possibility might change soon. Cutting edge rocket technology using nuclear-powered engines can accelerate spaceships today to a speed of three quarters of the speed of light after only one year of acceleration.

8.4 Time Reversal is Going Backwards in Time

Ah! Now, what you really have been waiting for is time reversal. Is it possible for time to reverse so that everyone, or the lucky one with the time machine, passes backwards in time? Philosophically speaking, you will have big problems when you convince your great-grandfather to marry a nicer girl. Yet, time reversal is also a popular topic (Boslough 1992, 208) (Bunch 1989, 215) (Coveney and Highfield 1990, 169) (Davies 1995 1996, 204) (Gardner 1964 1979, 257) (Klein 1988 1996, 185) (Sachs 1987) (Savitt 1995 1998, 12) (Shallis 1983, 63) (Sklar 1974 1976, 355) (Whitrow 1980, 321).

Our new ideas of Timespace and Universal Time suggest that time reversal is reasonable. Consider the following illustration in Figure 8.1. It expands Figure 3.1, our earlier illustration of speed in xyz S-space versus speed in Timespace:

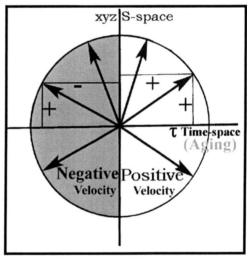

Figure 8.1 (G038B)
Negative Timespace Displacement

We know from experience that we can move in both the positive and negative xyz S-space directions. Can we then reverse aging by moving backwards along the fourth dimension of Timespace? As we reach the speed of light at the top of the circle in Figure 8.1, can we move into a space represented by the shaded area of the circle? Does time reverse in the shaded left half of the circle resulting in negative Timespace displacement?

Feynman refers to the direction, or arrow, of time using the laws of thermodynamics and entropy (Feynman 1965 1994, 102 and 107). This idea defines time by the concept of entropy. For example all processes, like mixing ink and water, proceed toward more chaos. This mixing process never reverses, even where ink separates from a water-ink mixture. Well-known Feynman diagrams do allow for time reversal (Davies 1995 1996, 205) (Reichenbach 1956 1999, 264) (Sachs 1987). Yet from a thermodynamic view, time reversal is not possible. Lawrence Sklar's conclusion in *Philosophy and Spacetime Physics* (Sklar 1985, 326) is that this thermodynamical model for time is "in doubt." Stephen Hawking also thinks that entropy, as a paradigm to understand time reversal, is not fruitful (Hawking 1988, 148). He argues that the universe is in a state of complete disorder at all times.

New possibilities for time reversal machines were presented by Kip Thorne in his book, *Black Holes & Time Warps* (Thorne 1994). His approach included the possibility of using closed timelike curves to travel back to the past (Hawking, Thorne, et Al. 2002, 36). Stephen Hawking's reaction to this idea is that time reversal via closed timelike curves has a quite small probability (Hawking, Thorne, et Al. 2002, 108).

8.5 Space as Small as a Singularity with Zero Volume

The academic subject of astronomy provides interesting applications for studying our new ideas of Timespace and Universal Time. In astronomy, you have extreme conditions of space. Space is smaller than the smallest subatomic particle and space is as great as the farthest galaxy. In astronomy, the black hole and the quasar represent these two extremes of space, where sizes and speeds require the ideas of four-dimensional space.

We will look down into the black hole with four-dimensional glasses. Our experience is comfortable with the idea of three-dimensional volume. In three-dimensional space concepts like "inside the room," "outside the room," "center of the room" and point in space are reasonable. In four-dimensional space these concepts are not reasonable. As discussed earlier, lower-dimension Euclidean projections from a higher-dimension space offer incomplete knowledge of the higher-dimension space. Concepts such as *three-dimensional* volume are not useful in space that is four-dimensional. With new concepts of Timespace and Universal Time, we understand that only four different three-dimensional worlds project down from four-dimensional space. We must not apply traditional concepts of time or xyz S-space to black holes and quasars. Exploring four-dimensional space near a black hole requires not only your two eyes, but also your brain.

Consider a black hole, the smallest of the small (Ray 1987, 141). The center of a black hole is a singularity. Our thinking is limited to just one of four projections down from four-dimensional space. Christopher Ray in *The Evolution of Physics* helps us to understand singularities:

"A singularity is a physical phenomenon which brings about a GI space-time. Loosely speaking, it is a 'place' where geodesics simply terminate."

He is saying that a singularity has no three-dimensional volume. A singularity, or black hole, is the condition where all matter in a very large star ceases to exist as a burning star. The matter collapses into a singularity, a "point" with no spatial volume. What conditions create a black hole? This collapse is an implosion followed by a dramatic explosion, called a supernova. Because the star no longer pushes outward with the energy of a nuclear furnace, the huge mutual gravitational attraction of the star's mass creates the inward collapse.

Remember, describing a singularity as a "point" in this discussion has the limitations associated with a three-dimensional xyz projection from four-dimensional space. The idea of all this mass located in zero three-dimensional volume seems absurd. Yet, four-dimensional geometry allows the non-existence of three-dimensional xyz S-space volume.

Theorists use the expression that "black holes have no hair." (Seeds 2001,

304) Once the matter, like our spaceship, enters the event horizon surrounding a back hole, it retains only three properties---mass, angular momentum, and electrical charge. The black hole "star" has no volume. The black hole has no density equaling mass divided by volume. The black hole has no temperature which requires vibration motion within a given three-dimensional volume.

With our Euclidean approach, using Timespace and Universal Time, we see that four-dimensional space has no use for traditional three-dimensional volume. The concept of volume is relevant to the xyz, three-dimensional projection from four-dimensional space.

8.6 Falling into a Black Hole

Using Einstein's theory, falling into a black hole provides another confusing paradox (Seeds 2001, 305). According to the theory of relativity, the spaceship accelerates as it falls. Eventually, the ship reaches maximum speed, the speed of light just above the singularity. We call this position the event horizon. The event horizon often is described as a sphere surrounding the singularity. This sphere's radius is equal to the Schwarzschild radius. Traditional physics suggests that any object with matter on this sphere can leave the black hole only if it has an escape velocity equal to the speed of light.

Using Einstein's theory creates a paradox as a spaceship gravitationally falls into a black hole. As the matter reaches the speed of light while it falls into the black hole, Einstein's theory states that observation outside of the falling spaceship watches traditional time stop. Clocks stop ticking. If time stops, does the spaceship stop falling or does the spaceship fall below the Schwarzschild radius at a constant, maximum speed of light? According to Einstein's theory nothing goes beyond the sphere of the event horizon.

The paradox is seen from observations inside the spaceship. Einstein's theory states that from inside the falling spaceship the spaceship will proceed toward the singularity. The ship moves past the event horizon as it descends to the singularity at the black hole's center. So, what really happens? Does the spaceship make it into the black hole, or not? Hopefully, you see that this paradox rests in the misuse of concepts limited to three-dimensional xyz S-space.

8.7 Timespace and Universal Time Explain Falling into a Black Hole

Timespace and Universal Time gives a better solution to objects entering into black holes. Einstein's theory does not have a good solution when relativity theory creates a paradox using two different observers.

Using simple calculus you can derive an equation for accelerations in Timespace. Start with Equation 8.1, formerly Equation 3.1, that relates velocities in Timespace τ and xyz S-space. This equation describes that the total speed in four-dimensional physical space is c, the speed of light. Differentiate Equation 8.1 with respect to Universal Time T to obtain Equation 8.2. Equation 8.2 shows the relationship between Timespace τ and xyz S-space accelerations. Note that these equations are limited to one xyz S-space projection from four-dimensional physical space.

$$v_\tau{}^2 + v_S{}^2 = c^2 \qquad \textbf{Equation 8.1}$$

$$a_\tau = -\frac{v_S}{v_\tau}a_S = -\left(\frac{c}{0}\right)a_S = -\infty \qquad \textbf{Equation 8.2}$$

Equation 8.2 also shows the conditions as the falling spaceship reaches the speed of light at the black hole's event horizon. If v_S equals c, then v_τ equals zero. See Figure 8.1. At this condition, according to Einstein's theory, time stops. Using our new ideas of Timespace and Universal Time, Equation 8.2 shows that when the spaceship reaches the speed of light at the event horizon of the black hole, the acceleration in Timespace will be negative infinity. At the black hole's event horizon, the spaceship's condition places the spaceship at the top of the circle in Figure 8.1. With an infinite acceleration to the left in Figure 8.1, the spaceship will enter the Schwarzschild radius sphere. Inside the event horizon Timespace displacement is negative. **The spaceship descending into a black hole moves past the black hole's event horizon sphere where time reverses direction!**

Equation 8.2 and Figure 8.1 show that velocities at quarter boundary conditions dominate Timespace accelerations. Regardless of the spaceship's acceleration in xyz S-space, velocities in xyz S-space and Timespace determine motion near and inside a black hole:

◆ If v_s equals zero and v_τ equals c or -c, then corresponding acceleration in Timespace, a_τ, equals zero.

◆ If v_s equals c or -c and v_τ equals zero, then corresponding accelerations in Timespace, a_τ, equal $-\infty$ or $+\infty$.

◆ Inside the black hole's spherical event horizon, matter oscillates back and forth between extreme velocities of c and -c.

These conditions for acceleration suggest that matter within the black hole's spherical event horizon would be trapped inside.

The experience of falling into a black hole is like reaching and transgressing an energy potential peak where there is no return. All of this occurs without the spaceship exceeding the speed of light. This result allows entering the event horizon and allows for time reversal in the form of negative Timespace displacement. **As Timespace displacement moves backwards, scalar Universal Time T stays consistent like the vibration of scalar Universal Frequency! Universal Time is a scalar with no quality of direction. Although tempting, thinking that Universal Time proceeds as usual in a forward direction is improper. Only Timespace displacement has a physical direction. Universal Time is a scalar and has no direction.** Within an event horizon is a world where matter oscillates within this sphere of capture. In simple language, our Euclidean geometry now suggests that "time reversal" is the mode of time displacement within the spherical event horizon of a black hole.

8.8 Escaping from a Black Hole

Consider the process of escaping from a black hole. We will use standard ideas for escape from the gravitational pull of a large massive star called a black hole. So, with our new ideas for Timespace and Universal Time, we must now realize our work is limited to one xyz S-space projection of a possible four projections from four-dimensional space.

Equation 8.3 is the traditional equation for the escape velocity necessary to leave the gravitational pull of mass M at radius r_{esc}. The term G is a constant number from Newtonian gravitation.

$$v_{esc} = \sqrt{\frac{2GM}{r_{esc}}}$$

Equation 8.3

If we place Equation 8.3 in Equation 8.1 for the velocity in xyz S-space, v_s, then Equation 8.4 for the speed in Timespace necessary to escape the mass M results:

$$v_\tau (Escape) = \sqrt{c^2 - \frac{2GM}{r_{esc}}}$$

Equation 8.4

The traditional equation for the Schwarzschild radius is Equation 8.5. This equation applies when the speed of any mass, like a spaceship, falls into mass M, like a black hole, through xyz S-space. As a spaceship falls into a black hole, the spaceship eventually reaches the speed of light at this radius. Equation 8.5 gives the traditional physics equation for this radius:

$$r_{Schwarzschild} = \frac{2GM}{c^2}$$

Equation 8.5

If the escape radius r_{esc} about mass M equals the Schwarzschild radius, an interesting result occurs. Combining Equations 8.4 and 8.5, we see that the speed through Timespace, v_τ, equals zero. In the language of Einstein's relativity theory: Time stops.

An agreement with general relativity also occurs. If the escape radius r_{esc} is greater than the Schwarzschild radius, Equation 8.4 equals the square root of a positive number. Escape can occur. According to the theory of general relativity, the speed of timekeeping, known as the speed through Timespace v_τ, increases with increasing values for the radius of escape, r_{esc}. Equation 8.4 shows this relationship.

More interestingly, if the escape radius r_{esc} is less than the Schwarzschild radius, Equation 8.4 becomes the square root of a negative number. In the language mathematics, the square root of a negative number is irrational. It is possible to interpret this irrational result. No escape from a black hole occurs if matter enters the spherical event horizon. Again, this is the traditional physics of a black hole.

So, once again we see how redefining time as Timespace and Universal Time provides Euclidean geometry solutions. These solutions for time and space compare well with traditional, complex solutions of non-Euclidean geometry.

Chapter 9
Comparison of Special Relativity with Timespace

*"Newton was the first to succeed in finding a clearly formulated
basis from which he deduced a wide field of phenomena
by means of mathematical thinking
--- logically, quantitatively, and in harmony with experience."*

--- Albert Einstein (Calaprice 1996, 074)

9.1 Timespace and Special Relativity

All equation derivations in Chapter Six include the reference frame condition of v_s equals zero, where Timespace τ equals $\tau_o = cT$. So, Chapter Six comparisons of Euclidean Timespace with Einstein's time dilation are comparisons within Einstein's special theory of relativity. The special theory o f relativity describes only motions with constant speed.

If we'd incorrectly used Einstein's time dilation, we would have specified $\tau_o = ct$. But, this equation would NOT be consistent with the use of Timespace τ and Universal Time T, where $\tau = v_\tau$ T. We must recognize that Einstein's time dilation is an idea created using Lorentz's invariant geometry and traditional time t. Einstein's time dilation does NOT use the Euclidean geometry of Timespace and Universal Time.

Chapters Seven and Eight demonstrate that Euclidean Timespace adequately describes motion of acceleration. The xyz projection from four-dimensional space is similar to Newtonian acceleration in three-dimensional xyz S-space. For example, Chapter Seven illustrates accelerations in the twin trip out and back along the x axis. Chapter Eight discusses acceleration as an object falls into a black hole. Einstein would categorize these accelerated motions as motion that his 1913 theory of general relativity describes. We discuss Einstein's general relativity and its concept of space in the next chapter.

9.2 Advantages of Timespace and Universal Time

Several advantages come from four-dimensional Euclidean descriptions of time and space. These advantages:

◆ These descriptions provide visual understanding in three-dimensional projections from four-dimensional space.

◆ Timespace vector displacement describes the quality of time that has past, present, and future. It allows a visual and mathematical description for "point in time."

◆ Timekeeping frequency, as well as Universal Time, provides a scalar time that Eastern philosophy describes time as "all present" with no beginning nor end.

◆ Scalar timekeeping frequency is the concept of time in wave motion.

◆ Universal Time has all of the qualities of Newtonian absolute time, except it can be universally used for the measurement of motion in four, or more, dimensions.

◆ Persons on Earth with only three-dimensional experience easily understand Euclidean geometry of four-dimensional space and time.

Timespace and Universal Time have one limitation: The Euclidean geometry approach to four-dimensional space does not completely describe four-dimensional space. The four different, three-dimensional projections from four-dimensional space explain the possible experience of time travel. Yet, these projections show that Euclidean geometry of Timespace and Universal Time can only provide a limited knowledge of four-dimensional, Euclidean space.

In comparison, Einstein's theories provide mathematically complete descriptions of non-Euclidean geometry. Non-Euclidean geometry, however, is more demanding than four-dimensional, Euclidean geometry. Non-Euclidean geometry requires a knowledge of higher mathematics. And beyond this, non-Euclidean geometry presents more difficult, non-visual geometry. Since humans' experience with space, so far, is three-dimensional, we find non-Euclidean geometry difficult to understand. We find easier to accept that when objects accelerate in xyz S-space, their total speed in four-dimensions is the speed of light at all times.

9.3 The Role of Invariance in Both Special Relativity and Timespace

Lorentz geometry created the use of algebraic invariants (Hilbert 1993, vii) (Taylor and Wheeler 1992, 8). Algebraic invariants provide the mathematical foundation for Einstein's special theory of relativity. Three examples of geometric invariance simplify and illustrate the idea of invariance:

(1) **The diameter of a circle**. See Equation 9.1 and Figure 9.1. Connect any point on a semicircle to both end points of the semicircle diameter. For any point on the semicircle, this connection creates a right angle of 90 degrees (Wolf 1988, 121). Via the Pythagorean Theorem, corresponding values of connections a and b produce the same circle diameter d. This example illustrates geometric invariance. This example, however, has no known application for describing space and time.

$$a^2 + b^2 = d^2 \qquad \text{Equation 9.1}$$

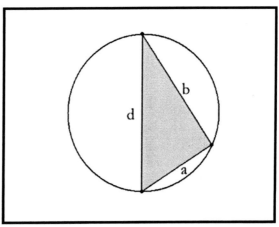

Figure 9.1 (G065f)
Circle Diameter Invariance

(2) **Non-Euclidean Space or Lorentz Geometry.** Before we illustrate invariance in four dimensions, first consider the circle in Figure 9.2. The geometry of this circle can be represented with Equation 9.2:

$$x^2 + y^2 = (ct)^2$$ **Equation 9.2**

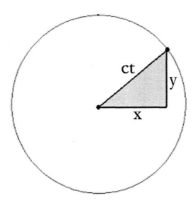

Figure 9.2 (G066f)
Two-Dimensional Lorentz Invariance

Here coordinates x and y combine and equal the circle's radius, ct. This geometry is adequate for describing a full plane of space when different values of t are possible creating larger and larger circles. The value for c remains constant for all possible circles. This is an example of the geometric invariance of the metric, or distance, ct.

Now we apply Figure 9.2 to three-dimensional xyz S-space. Equation 9.3 and Figure 9.3 represent a spherical light beam expanding outward into three-dimensional space:

$$x^2 + y^2 + z^2 = (ct)^2 \qquad \text{Equation 9.3}$$

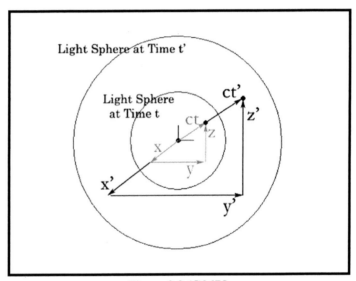

Figure 9.3 (G067f)
Three-Dimensional Lorentz Invariance

Motion of x', y', z', t' Produces
Four-Dimensional, Non-Euclidean Geometry

Equation 9.3, Lorentz's invariant equation and Einstein's equation for special relativity, can describe all possible locations in three-dimensional space for varying values of time t. Figure 9.3 requires only a simple Euclidean description of space and a Newtonian description of absolute time.

If, however, the reference frame, x', y', z', moves at speed v, then we find that a description of this non-Euclidean space must account for a unique value of time t along with each value of x, y and z. Why? A unique value of time t for each x-y-z point in space secures that each reference frame will measure the same value for the speed of light. Euclidean geometry does not apply to geometry of moving reference frames like non-Euclidean geometry. In either two or three dimensions, Euclidean geometry applies to geometric figures which are at rest (Kopff 1923, 48). In contrast, non-Euclidean geometry is geometry in motion.

Non-Euclidean geometry in the special theory of relativity requires each position in space to have a fourth and unique value for time t. Space becomes xyz-t. Authors often compare this "four-dimensional" universe to the surface of a sphere. In relativity theory, the nature of Minkowski's four-dimensional space-time geometry is analogous to the surface of a sphere. The surface of a sphere is "warped" two-dimensional Euclidean space. There is no center on this spherical surface. The surface of a sphere is available for infinitely long travel. Concepts like "beyond" has no meaning for travel on the surface of a sphere. On a sphere one returns to the same location by moving away from this location via a "straight" line.

Einstein in his 1905 paper on special relativity chose Equation 9.3 as the representation of a spherical wave in the context of Euclidean geometry. This light wave leaves a source at the origin of two coordinate systems (Miller 1981, 403). Analysis of these two coordinate systems observed in relative motion produces Einstein's equation for time dilation, Equation 1.1.

So, Einstein's time dilation equation gives the relationship between t and t' when the reference frame of x', y' and z' moves at speed v. Einstein's non-Euclidean approach is geometry where time is the fourth variable. According to the special theory of relativity, Einstein's geometry represents a spherical wave of light leaving an event. This spherical wave carries the information of this event. We must realize that the origin of relativity's non-Euclidean geometry comes from the context of a spherical light wave traveling in three-dimensional space.

Simultaneity arguments by Einstein play an important role in his theory. For Einstein's relativity, space-time intervals exist between events, just as distance intervals exist between points in three-dimensional space.

Minkowski's 1908 idea that time is the fourth, non-Euclidean dimension is complex and confusing. Minkowski's idea is a tangential thought to that of Einstein's special relativity.

Minkowski introduces us to the idea of a four-dimensional continuum where he says space and time can now be considered a "space-time continuum." This continuum terminology masks that space-time in the special theory of relativity never use Euclidean geometry. Minkowski's terminology also masks that space-time is a four-dimensional geometry that is not Euclidean geometry. Einstein tried to eliminate confusion by emphasizing that the general theory of relativity is not a Euclidean continuum (Einstein 1920 1921, 93).

Invariance in Einstein's special theory of relativity implies that space has three dimensions. Yet, invariance leads us to space-time geometry with four dimensions. (Hilbert 1993) (Miller 1977 1981, 195) (Murchie 1961, 547) (Taylor and Wheeler 1963, 8)

(3) **Euclidean four-dimensional space.** Eddington warns that there is no suggestion that the extra fourth dimension is anything but a fictitious construct, useful for representing a property found in surfaces (Eddington 1933, 43). Our presentation introduces Euclidean four-dimensional physical space. The fourth dimension in Euclidean geometry is the physical dimension we call Timespace.

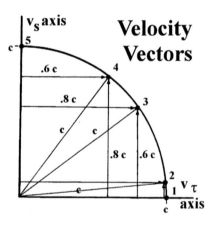

Figure 9.4
Euclidean Invariance

From any point on a quarter circle, draw a line perpendicular to the radius of a quarter circle. These lines produce right triangles shown in Figure 9.4. See Equation 9.4:

$$v_s^{\,2} + v_\tau^{\,2} = c^2 \qquad \text{Equation 9.4}$$

For the quarter circle in Figure 9.4, the radius of the circle always equals c, the speed of light. The geometry of the quarter circle immediately illustrates that the speed of light is constant. The radius of the quarter circle represents geometric invariance. Numbered positions on the quarter circle progress from one to five. These five positions illustrate motion accelerating from a speed of zero to a value of c, the speed of light. With a speed of zero in xyz S-space at position one, motion in Timespace is maximized at the speed of light. With motion maximized at the speed of light in xyz S-space at position five, motion in Timespace is zero. For Einstein's relative motion at the speed of light at position five, Einstein says that time stops at this fifth position.

Now we can see redefining time as Timespace and Universal Time along with Equation 9.4 provides a simple Euclidean view of four-dimensional space. Motion through the fourth dimension varies. This variation in timekeeping is speed through Timespace τ.

Universal Time goes beyond the concept of Newton's absolute time where time connotes the passage of time in three-dimensions of space. Universal Time is a scalar with no beginning and no end. Universal Time does not stop, even though Timespace displacement might stop under certain conditions.

Using Universal Time and the contraction of space for moving objects, this Euclidean view provides for light's speed to always measure with the same value. In four-dimensional Euclidean geometry, the motion of any object is equal to c, the speed of light.

9.4 Why Does Einstein's Relativity Choose No Motion for the Observer?

C. D. Broad's statement in Chapter Two attacks the philosophical validity of any physics that takes all motion to be relative (Broad 1914, 277). He states that you can talk about neither the direction nor the speed of a body if all motion is relative to the observer. As we saw with the fly in a moving automobile, the condition of constant speed is identical to zero speed. And, with anyone at rest on the moving Earth, that person's speed can be either zero or 1000 mph. C. D. Broad's criticism appears valid and leads one to think that any arbitrary speed and direction can be assigned to motion by selecting the appropriate reference frame.

Why does Einstein choose the speed of the observer's reference frame to be zero? He uses the symbolism t_0 and l_0 to represent local measurements at a zero speed. Einstein states that we observe time dilation and length contraction from a reference frame at rest, or without motion. Einstein's choice of zero speed for the observer seems to contradict the postulate of Einstein's 1905 special relativity. The symbols t_0 and l_0 stand for time and length measured at rest. Yet, Einstein's postulate states that no absolute rest exists.

We can turn to the ideas of Timespace and Universal Time that use Euclidean geometry for an answer to why Einstein's observer's reference frame's speed must be zero. As in Equation 9.4, the total of all speeds in four-dimensional Euclidean space is the speed of light. **If we choose any speed other than zero for the Universal Reference Frame or Einstein's observer, then the total of all speeds in four-dimensions exceeds the speed of light.** This is a violation of the principle that the speed of light is the maximum speed obtainable. This fundamental idea prevents Einstein's local observer's speed from being anything other than zero. Likewise in the case of Timespace and Universal Time the speed of a Universal Reference Frame is necessarily zero. In a Universal Reference Frame observers do not perceive their motion and they consistently measure identical Universal Time T everywhere in the universe. **The process of "observing" time dilation is inherent to Einstein's method and is the fundamental source of paradoxes in Einstein's theory.**

Timespace and Universal Time as presented in this book do not require "observing" time dilation for clocks in relative motion. In Chapter Seven the Euclidean approach to the twin trip within the x-y-τ projection eliminates the paradox of conflicting timekeeping rates. Universal Time is a measure of scalar time in traditional units of minutes and hours on clocks. Everywhere, Universal Reference Frames provide for the identical measurement of Universal Time. Only Timespace displacement varies with varying speed through Timespace. Timespace displacement is a measure of distance on moving clocks in units of meters, kilometers and light years.

Chapter 10
Comparison with General Relativity

"Now I'm opening out like the largest telescope that ever was!
Good-bye feet!" (for when she looked down at her feet,
they seemed to be almost out of sight, they were getting so far off)

--- from *Alice in Wonderland* by Lewis Carroll

10.1 Timespace and General Relativity

General relativity accounts for variations in the rate of timekeeping in accelerated clocks where force causes the acceleration. Force and acceleration exist in a three-dimensional projection from four-dimensional Euclidean space. Traditional physics is in the context of xyz S-space.

In 2001, Sir Martin Rees, Astronomer Royal in Great Britain, concludes in *Our Cosmic Habitat* that force does not exist in higher, four-dimensional space (Rees 2001, 148):

> "It is actually because there are three space dimensions that electric and gravitational forces obey an inverse square law. This dependence is easiest to appreciate in terms of Faraday's concept of lines of force. A shell of radius r around a mass or charge has an area proportional to r^2; the force falls off as $1/r^2$ because at larger radii the lines of force are spread over a bigger area and their effect is diluted. If there were a fourth spatial dimension, the area of a sphere would be proportional to r^3 instead of r^2, and the force would follow an inverse cube law. ... Things would be catastrophically different if, instead, gravity obeyed an inverse cube law: a planet that was slightly slowed down would plunge into the Sun; and if it were slightly speeded up, it would spiral outward toward cold interstellar darkness."

Our approach here reaches this same conclusion. We can understand that force, as it exists in xyz S-space, does not exist in four-dimensional space. Consider the basic idea that the total speed of all objects at all times equals the speed of light. Speed in four-dimensional space is universal motion at all times. The traditional idea of force exists in only one of four three-dimensional projections, that of xyz S-space. Force does not exist in four-dimensional space, where everything at all times moves at a constant speed of light.

Einstein's general theory of relativity uses eloquent, higher mathematics of non-Euclidean geometry. Einstein based his theory upon three-dimensional geometry of space and four-dimensional geometry of the space-time continuum. The theory's unique contributions are that three-dimensional space "warps" about objects with mass. The warping of space about objects with mass explains, or replaces, gravitation force and acceleration. Today authors describe four-dimensional space as warped, three-dimensional space.

10.2 Einstein's Theories of Relativity Use Complex Mathematics

In four-dimensional space, ideas like volume, inside and outside are invalid. Einstein recognizes limitations on concepts such as volume, inside and outside. Everyday experience gives us experience with volume, inside and outside. These experiential ideas only apply to the xyz S-space projection from four-dimensional space.

In contrast, Einstein's four-dimensional field equations use four-dimensional matrices. These matrices describe non-Euclidean space. Yet Einstein's matrices are a mathematical description of four-dimensional space. These matrices are invisible. Without four-dimensional experience or higher mathematics knowledge, we find Einstein's explanation inaccessible.

A short story illustrates the need for higher mathematics knowledge to understand relativity theory. Einstein allegedly told this story when someone asked him to explain his theory without mathematics. Einstein's story went something like this: A blind man asked a friend, "What is milk?" The friend answered: "Milk is a white liquid." The blind man knew what a liquid was, but he asked, "What is white?" The friend answered: "White is the color of swan feathers." The blind man knew what feathers were, but he asked, "What is a swan?" Answer: "A swan is a bird with a crooked neck." The blind man knew what a neck was, but asked: "What do you mean by crooked?" Not seeing an end to the blind man's questions, his friend grew frustrated and grabbed the arm of the blind man and bent it. "This is crooked." With some contemplation the blind man shouted: "Now I know what milk is!" So, without mathematics, even if someone twists your arm to make you say you understand, you actually do not understand anything at all.

Fritjoy Capra's 1976 book, *The Tao of Physics*, is a very popular attempt to describe physics without mathematics. Abram Pias, distinguished biographer and colleague of Einstein, wrote on the title page of his personal copy of Capra's book. See Figure 10.1 (Capra 1984):

Figure 10.1 (G081f)
Handwriting of Abram Pias

10.3 Space Surrounding Objects of Mass and Einstein's Warped Space

Einstein's theory of general relativity describes how the rate of timekeeping varies near a mass M, like a planet. This variation in timekeeping rates is Einstein's idea of warped three-dimensional space about any mass M. This idea of warped three-dimensional non-Euclidean space, provides a different approach than that of Newtonian universal gravity. This theory predicted that light coming to Earth would be bent from a straight path. Einstein's relativity says warped space near a large mass like the Sun causes light to bend. This bending is just like any spaceship's path would bend by the pull of Newtonian gravity. Many complicated experiments attempted to verify this prediction of the theory of general relativity. The experiments to verify general relativity have produced varying results that generally support Einstein's general theory (Will 1986, 65).

Since our Euclidean approach to time and space includes a variable for the rate of timekeeping, v_τ, we can construct a simple general equation for the rate of timekeeping about an object of mass M. Traditional physics operates in one xyz S-space projection from four-dimensional space. Equation 10.1 gives the traditional expression for Newtonian gravitational acceleration g:

$$g = a_s = \frac{GM}{r^2} \qquad \textbf{Equation 10.1}$$

To obtain an equation for varying rates of timekeeping about a mass M, combine three equations to eliminate a_τ:

◆ (1) Equation 8.1 states the total speed in four-dimensional Euclidean space is the speed of light.
◆ (2) Equation 8.2 relates speeds and accelerations in Timespace.
◆ (3) Equation 10.1 presents the expression for Newtonian gravitational acceleration.

Combine these three equations.

◆ Equation 10.2 results. Equation 10.2 is a general expression for time-keeping rate v_τ about mass M at distance r from the mass's center:

$$v_\tau = \sqrt{c^2 - \frac{v_s^2 a_s^2 r^4}{G^2 M^2}} \qquad \textbf{Equation 10.2}$$

The following qualities of Equation 10.2 represent its value:

◆ **All clocks in the universe are in a Universal Reference Frame and measure Universal Time T**. The traditional physics understanding of Equation 10.2 is limited to the xyz S-space projection from four-dimensional Euclidean space. Three other projections exist.

◆ With Equation 10.2 we conclude that the timekeeping rate would be the expected maximum value of the speed of light when near mass M and the clock is at rest. If v_s equals zero, then v_r equals c, the speed of light.

◆ With Equation 10.2 we conclude that when any object is infinitely far away from a mass M where the acceleration due to gravity is zero, the timekeeping rate is the expected maximum value of the speed of light . If a_s equals zero, then v_r equals the speed of light.

◆ Equation 10.2 shows that for any finite speed value in xyz S-space, where $0<v_s<c$, and for any finite value for acceleration a_s, timekeeping rate varies with r, the distance from the center of mass M.

◆ Equation 10.2 includes r to the fourth power. So, r dominates. Equation 10.2 implies that as the distance from mass M increases, the observed timekeeping rate of more distant clocks would decrease accordingly. This is in agreement with Einstein's general theory of relativity.

◆ Equation 10.2 is the general form of Chapter Eight's Equation 8.4. Equation 8.4 describes the timekeeping rate on clocks surrounding a black hole. This can be shown with substitution of the Equations 8.2 and 8.3 into Equation 10.2.

10.4 Universal Equivalence Principle Replaces Einstein's Equivalence Principle

Einstein's Equivalence Principle is another example of an idea creating a paradox. Universal Reference Frames need not be inertial reference frames defined by Newton's first law of motion. Einstein says a person free-falling in an elevator is weightless as if in space far distant from any planetary mass and its gravitation. Einstein's equivalence principle considers this free falling condition as locally equivalent to an inertial reference frame. Is the free-falling person weightless or is this person accelerating with the force of gravity? Which condition is true? According to Einstein's theory, the answer is both are true.

Universal Time T and Universal Reference Frames eliminate this paradox. A Universal Equivalence Principle applies with equal validity to constant speed motion. The story of the fly in the automobile illustrates this Universal Equivalence Principle. A reference frame at rest is equivalent to a reference frame moving at a constant 50-mph speed.

In fact, while the auto accelerates as the fly sits buckled up in its car seat, the fly also experiences a Universal Reference Frame and locally measures Universal Time T. **You can perceive force and its acceleration in a Universal Reference Frame. But, you can't perceive the Universal Reference Frame's motion.** Clocks in inertial reference frames moving with constant speed as well as accelerated clocks measure Universal Time T. The nature of Universal Time is truly universal. **A Universal Principle of Equivalence universally applies. A Universal Principle of Equivalence replaces Einstein's equivalence principle for gravitational accelerated motion.**

10.5 Einstein States that No Empty Space Exists

Many authors write about space. Often they show experiential bias toward visualizing four-dimensional space in three-dimensional illustrations. We hope that the ideas here help your thinking about four-dimensional space. Physical space is any multidimensional space where motion occurs. With this Euclidean approach, higher-dimensional space such as eleven-dimensional superstring theory is valid.

Near the end of Einstein's life, Einstein was consistently concerned about the idea that no empty space exists. In 1952, Einstein added a note to the preface of the fifteenth edition of *Relativity: The Special and the General Theory* (Einstein 1961, vi):

> "I wished to show that space-time is not necessarily something to which one can ascribe a separate existence, independently of the actual objects of physical reality. Physical objects are not *in space*, but these objects are *spatially extended*. In this way the concept of "empty space" loses its meaning."

In the text of this book Einstein again refers to empty space losing its meaning (Einstein 1961, 151):

> "Thus even this theory does not dispel Descartes' uneasiness concerning the independent, or indeed, the *a priori* existence of "empty space." The real aim of the elementary discussion given here is to show to what extent these doubts are overcome by the general theory of relativity."

Again in 1953, in the forward to *Concepts of Space: The History of the Theories of Space in Physics,* by Max Jammer, Einstein wrote that empty space does not exist (Jammer 1954 1957, xv) (Jammer 1954 1993, xvii):

> "The victory over the concept of absolute space or over that of the inertial system became possible only because the concept of the material object was gradually replaced as the fundamental concept of physics by that of the field. Under the influence of the ideas of Faraday and Maxwell the notion developed that the whole of physical reality could perhaps be represented as a field whose components depend on four space-time parameters. If the laws of this field are in general covariant, that is, are not dependent on a particular choice of coördinate system, then the introduction of an independent (absolute) space is no longer necessary. That which constitutes the spatial character of reality is then simply the four-dimensionality of the field. There is then no "empty" space, that is, there is no space without a field."

Perhaps these statements by Einstein explain why in 2005 three books on multiple dimensions of space chose not to define space in their glossaries (Krauss 2005, 257)(Kaku 2005, 381)(Randall 2005, 459). As recently as 2006 Max Jammer, an authority on the history of space concepts, concludes that time differs from space (Jammer 2006, 300).

Empty three-dimensional space exists like the number "3" that you learned as a child. Empty three-dimensional space is nothing more than Euclid's geometry. Here Einstein asks you to replace this three-dimensional space with the *four-dimensionality* of a force field.

Einstein's four-dimensional field as four-dimensional space is not the Euclidean geometry of Timespace and Universal Time presented here. Timespace and Universal Time use Euclidean four-dimensional geometry. Four different three-dimensional projections from four-dimensional space describe this non-visible space.

10.6 Timespace and Universal Time Can Describe Accelerated Motion Like General Relativity

The recently stated need for a new definition for time brings us back to a reasonable approach to understanding time as well as space. This understanding redefines time as Timespace and Universal Time. This four-dimensional geometry uses Euclid's geometry and the idea of empty space. Further, this four-dimensional Euclidean geometry allows for Universal Reference Frames in xyz S-space. As motion in xyz S-space occurs, corresponding motion in Timespace τ produces rates of timekeeping, v_τ, that vary with acceleration and a corresponding vectorial displacement in Timespace τ. These motions, however, exist in four different three-dimensional projections from four-dimensional space. And, like the non-Euclidean geometry of Einstein Relativity, these three-dimensional projections do NOT provide a visual description of four-dimensional space.

Has a unique Euclidean solution been considered before this presentation? The answer is perhaps yes. In 1840 Nicholas Lobachevski, Russian Professor of Mathematics in the University of Kasen, authored a book, *The Theory of Parallels* (Lobacheveski 1840 1914). Lobachevski was the first person ever to publish about non-Euclidean geometry (Lobacheveski 1840 1914, 3) (Sommerville 1914 1958, 20) (Vasiliev 1924, 95).

George Bruce Halsted translated Lobachevski's *The Theory of Parallels* into English in 1891. This translation was published in 1914. On the last page of this 1914 translation, Halsted commented on the use of geometry in Einstein's special theory of relativity (Lobacheveski 1840 1914, 50):

"I postulated that the phenomena (of relativity) happened in a Lobachevski space, and reached by very simple geometric deduction the formulas of the relativity theory. Assuming non-euclidean [sic] terminology, the formulas of the relativity theory not only essentially simplified, but capable of a geometric interpretation wholly analogous to the interpretation of the classic theory in the euclidean [sic] geometry."

I have no evidence that Halsted published details of his Euclidean geometry for space and time. Another interpretation of Halsted's statement is that Euclidean geometry and non-Euclidean geometry are equivalent. Our new ideas here testify they are not equivalent.

References

Abbott and Burger, 1880 1994, *Flatland and Sphereland*, (New York: Harper Perennial).

Aczel, Amir, 1999, *God's Equation*, (New York: Four Walls Eight Windows).

Adler, Kuper, Rosner & Weil, Editors, 1990, *Developments in General Relativity, Astrophysics and Quantum Theory*, Rosen Jubilee, (Bristol: I. O. P. Publishing).

Ahlfors, Lars and Sario, Leo, 1960, *Riemann Surfaces*, (Princeton: Princeton University Press).

Albert, D., 2000, *Time and Chance*, (Cambridge, MA: Harvard University Press).

Alexander, H. G., 1959, "The Paradoxes of Confirmation—A Reply to Dr. Agassi," *The British Journal for the Philosophy of Science*, (Discussions), 10 (November): 229-234.

Alexander, Editor, 1956, *The Leibniz-Clarke Correspondence*, (Manchester, England: Manchester University Press).

Alexander, S., 1920 1950, *Space, Time and Deity*, Two Volumes, (New York: The Humanities Press).

Apfel, Necia H., 1985, *It's All Relative*, (New York: Lothrop, Lee & Shepard).

Aristotle's Physics, 1969, Translated by Hippocrates G. Apostle, (Bloomington: Indiana University Press).

Arthur, James, 1909, *Time and its Measurement*, Reprint, (Chicago: Popular Mechanics Magazine).

Arzelies, Henri, 1972, *Relativistic Point Dynamics*, (Oxford: Pergamon Press).

Asimov, et. Al., 1946, *Adventures in Time and Space*, (New York: Ballantine Books Random House).

Asimov, Isaac, 1964, *Adding a Dimension*, (New York: Discus Books by Avon).

Asimov, Isaac, 1964 1972, *Biographical Encyclopedia of Science and Technology*, (New York: Doubleday & Company).

Asimov, Isaac, 1968, The Fourth Dimension, *Science Digest*, March, 80-81.

Asimov, Janet, 1988, *The Package in Hyperspace*, (New York: Walker and Co.).

Audoin, C. and Guinot, B., 1998 2001, *The Measurement of Time: Time, Frequency and the Atomic Clock*, (Cambridge, England: Cambridge University Press).

Aveni, Anthony, 2002, *Empires of Time: Calendars, Clocks, and Cultures*, Revised Edition, (New York: University Press of Colorado).

Banchoff, Thomas, 1990 1996, *Beyond the Third Dimension*, (New York: Scientific American Library).

Barbour, Julian, 1999 2000, *The End of Time*: The Next Revolution in Physics, (Oxford: Oxford University Press).

Barnett, Jo Ellen, 1998, *Time's Pendulum*, (New York: Plenum Press).

Barnett, Lincoln, 1948 1968, *The Universe and Dr. Einstein*, (New York: Bantam Science and Mathematics).

Barrett, Clifford, 1935 1956, *Philosophy*, (New York: Macmillan Co.).

Barrow, J., 1999 2000, *Between Inner Space and Outer Space*, (Oxford: Oxford University Press).

Barrow, John, 1994, *Origin of the Universe*, (New York: Basic Books).

Basri, Saul A., 1965, "Operational Foundation of Einstein's General Theory of Relativity," *Reviews of Modern Physics*, 37 (April): 288-315.

Bath, Geoffrey, Editor, 1980, *The State of the Universe*, (Oxford: Oxford University Press).

Beck, A. and Havas, P., 1987, *Collected Papers of Albert Einstein: Translations for Volume 1*, (Princeton: Princeton University Press).

Beck, A., Translator, 1995, *Collected Papers of Albert Einstein, Volume 5, The Swiss Years, 1902-1914*, (Princeton: Princeton University Press).

Becker, Richard, 1930 1967, *Electromagnetic Fields and Interactions: Volume I, Electromagnetic Theory and Relativity*, (New York: Ginn and Com-pany).

Begelman, M. and Rees, M., 1996 1998, *Gravity's Fatal Attraction*, (New York: Scientific American Library).

Begley, Sharon, 2003, Physicists Are Looking at How We Might Take a Trip through Time, *Wall Street Journal*, November 21.

Bell, E. T., 1937 1965, *Men of Mathematics*, (New York: Simon and Schuster).

Bell, Robert, 1910 1920, *Coordinate Geometry of Three Dimensions*, Second Edition, (London: Macmillan).

Berberian, Sterling K, 1961 1999, *Introduction to Hilbert Space*, (New York: AMS Chelsea Publishing).

Bergmann, Peter, 1942 1976, *Introduction to the Theory of Relativity*, (New York: Dover Publications).

Bernal, J. D., 1972 1977, *A History of Classical Physics*, (New York: Barnes & Noble).

Bernstein, J. and Feinberg, G., Editors, 1986, *Cosmological Constants*, (New York: Columbia University Press).

Bernstein, Jeremy, 1973 1980, *Einstein*, (Middlesex, U K: Penguin Books Ltd.).

Bird, J. Malcolm, Editor, 1921, *Einstein's Theory of Relativity and Gravity - Higgins Prize Essays*, (New York: American Scientific Publishing).

Bjorken, J. and Drell, S., 1964, *Relativistic Quantum Mechanics*, (New York: McGraw Hill).

Bjorken, James & Drell, Sidney, 1965, *Relativistic Quantum Fields*, (New York: Mcgraw Hill).

Bodanis, David, 2000, *E=mc²: A Biography of the World's Most Famous Equation*, (New York: Walker & Company).

Bogoliubov, 1969, *Problems of Theoretical Physics*, (Moscow, Russia: Nauka).

Bogoliubov & Shirkov, 1983, *Quantum Fields*, (Reading, M A: Benjamin-Cummings Publishing).

Bohm, David, 1965 1996, *Special Theory of Relativity*, (New York: Routledge Press).

Bohr, Neils, 1987, *The Philosophical Writings of Niels Bohr*, Volume I, (Woodbridge, C T: Ox Bow Press).

Bohr, Neils, 1987, *The Philosophical Writings of Niels Bohr*, Volume II, (Woodbridge, C T: Ox Bow Press).

Bohr, Neils, 1987, *The Philosophical Writings of Niels Bohr*, Volume III, (Woodbridge, C T: Ox Bow Press).

Bondi, Hermann, 1964 1980, *Relativity and Common Sense*, (New York: Dover Publications).

Bonola, Roberto, 1912 1955, *Non-Euclidean Geometry: Translated by Carslaw, H. S.*, (New York: Dover Publications).

Born, Max, 1969, Albert Einstein, Hedwig, und Max Born. *Briefweschsel*, 1916-1955, (Munich, Germany: Nymphenburger).

Born, Max, 1924 1962, *Einstein's Theory of Relativity*, (New York: Dover).

Born, Max, 1971, *The Born-Einstein Letters*, (New York: Walker and Company).

Boslough, John, 1985 1989, *Stephen Hawking's Universe*, (New York: Avon Books).

Boslough, John, 1992, *Masters of Time*, (New York: Addison-Wesley Publishing).

Boyer, Carl, 1968, *A History of Mathematics*, (New York: John Wiley & Sons).

Boyer, Carl, 1968 1991, *A History of Mathematics*, Second Edition, (New York: John Wiley & Sons).

Bragdon, Claude, 1913, *A Primer of Higher Space*, (New York: The Manas Press).

Bragdon, Claude, 1916 1925, *Four-Dimensional Vistas*, Second Edition, (New York: Knopf).

Brian, Denis, 1996, *Einstein a Life*, (New York: John Wiley & Sons).

Bridgman, P. W., 1927 1932, *The Logic of Modern Physics*, (New York: The Macmillan Company).

Bridgman, Percy, 1983, *A Sophisticate's Primer of Relativity*, Second Edition, (Middletown, CT: Wesleyan University Press).

Brisson, David W. Editor, 1977 1978, *Hypergraphics: Visualizing Complex Relationships in Art, Science & Technology*, (New York: AAAS Symposium 24).

Broad, C. D., 1914, *Perception, Physics, & Reality*, (Cambridge: Cambridge University Press).

Broad, C. D., 1925, *The Metaphysical Foundations of Modern Physical Science*, (New York: Harcout, Brace, & Company).

Broad, C. D., 1933, *Examination of McTaggart's Philosophy*, Volume 1, (Cambridge: Cambridge University Press).

Broad, C. D., 1938, *Examination of McTaggart's Philosophy*, Volume 2, (Cambridge: Cambridge University Press).

Broad, C. D., 1952, *Ethics and the History of Philosophy*, (New York: International Library of Psychology., Philosophy. & Scientific Method).

Bronowski, J., 1955, *The Common Sense of Science*, (Cambridge, MA: Harvard University Press).

Bruce, Colin, 1997, *The Einstein Paradox*, (Reading, MA: Perseus Books).

Bunch, Brian, 1989, *Reality's Mirror: Exploring the Mathematics of Symmetry*, (New York: John Wiley & Sons).

Burke, William L, 1980, *Space-time, Geometry, Cosmology*, (Millvally, CA: University Science Books).

Burtt, E A, 1924 1954, *The Metaphysical Foundations of Modern Physical Science*, Revised 1931 Edition, (New York: Doubleday Anchor Books).

Calaprice, Alice, Editor, 1996, *The Quotable Einstein*, (Princeton: Princeton University Press).

Calder, Nigel, 1979, *Einstein's Universe*, (London: British Broadcasting Corporation).

Calder, Nigel, 1979 1980, *Einstein's Universe*, (New York: Penguin Books).

Calder, Nigel, 1983, *Timescale: An Atlas of the Fourth Dimension*, (New York: The Viking Press).

Callahan, J. J., 1931, *Euclid or Einstein*, (New York: The Devin-Adair Company).

Campbell, Norman Robert, 1920 1957, *Foundations of Science: The Philosophy of Theory and Experiment*, (New York: Dover Publications).

Capra, Fritjof, 1984, *The Tao of Physics*, Second Edition., (New York: Bantam Books).

Carmichael, Robert, 1913, *The Theory of Relativity*, (New York: John Wiley & Sons).

Carnap, Rudolf, 1954 1958, *Introduction to Symbolic Logic and its Applications*, (New York: Dover Publications).

Carnap, Rudolf, 1966, *Philosophical Foundations of Physics*, (New York: Basic Books).

Carr, H. W., 1922, *The General Principle of Relativity*, (London: Macmillan and Co.).

Carr, Herbert Wildon, 1929 1960, *Leibniz*, (London: Dover Publications).

Carroll, Sean, 2004, *Spacetime and Geometry: An Introduction to General Relativity*, (New York: Addison-Wesley Publishing Company).

Carslaw, H. S., 1906 1912, *Non-Euclidean Geometry*, (Chicago: Open Court Publishing).

Cartan, Elie, 1922, *Lecons Sur Les Invariants Integraux*, (Paris: Librairie Scientifique).

Cartan, Elie, 1932, *Le Parallelisme Absolut et La Theorie Unitaire Du Champ*, (Paris: Hermann & C).

Cartan, Elie, 1932 1979, *Albert Einstein Letters on Absolute Parallelism 1929-1932*, (Princeton: Princeton University Press).

Cartan, Elie, 1934, *Les Espaces De Finsler*, (Paris: Hermann & Co.).

Cartan, Elie, 1938, *Lecons Sur La Theorie Des Spineurs*, Volume 1, (Paris: Hermann & Co.).

Cartan, Elie, 1938, *Lecons Sur La Theorie Des Spineurs*, Volume 2, (Paris: Hermann & Co.).

Carter, Edward, 1993, *One Grand Pursuit: A Brief History of the American Philosophical Society's First 250 Years*, (Philadelphia: American Philosophical Society).

Cartwright, Nancy, 1983, *How the Laws of Physics Lie*, (New York: Oxford: Clarendon Press).

Casimir, H., 1983, *Haphazard Reality*, (New York: Harper & Row).

Cassirer, Ernst, 1923 1952, *Substance and Function: Einstein's Theory of Relativity*, (New York: Dover Publications).

Cassirer, Ernst, 1929 1977, *The Philosophy of Symbolic Forms*, Volume 3, (New Haven, C T: Yale University Press).

Chaisson, Eric, 1988 1990, *Relatively Speaking: Relativity, Black Holes, and the Fate of the Universe*, (New York: W. W. Norton & Company).

Chandrasekhar, S., 1992, *The Mathematical Theory of Black Holes*, (Oxford: Oxford University Press).

Christodoulou, D. and Klainerman, S., 1993, *The Global Nonlinear Stability of the Minkowski Space*, (Princeton: Princeton University Press).

Clark, D. N., Editor, 1981, *Contributions to Analysis and Geometry*, (Baltimore, MD: The John Hopkins University Press).

Clark, Ronald, 1984, Einstein, *The Life & Times: An Illustrated Biography*, (New York: Harry N. Abrams, Inc.).

Cline, Barbara, 1965 1987, *Men Who Made a New Physics*, (Chicago: University of Chicago Press).

Cohen, R. and Seeger, R., Editors, 1970, *Ernst Mach: Physicist and Philosopher*, (Dordrecht, Holland: D. Reidel Publishing).

Cohn, Harvey, 1967 1980, *Conformal Mapping on Riemann Surfaces*, (New York: Dover Publications).

Cole, K. C., 1998, *The Universe and the Teacup: The Mathematics of Truth and Beauty*, (New York: Harcourt Brace & Company).

Cole, K., 2001, *The Hole in the Universe*, (New York: Harvest Book, Harcourt, Inc.).

Coleman, James, 1954 1972, *Relativity for the Layman*, (New York: Penguin Books).

Copi, Irving M., 1967, *Symbolic Logic*, Third Edition, (New York: Macmillan Company).

Courant, R., 1924 1953, *Methods of Mathematical Physics*, Volume 1, (New York: Interscience Publishers, Wiley).

Courant, R., 1924 1966, *Methods of Mathematical Physics*, Volume 2, (New York: Interscience Publishers, Wiley).

Coveney, P., and Highfield, R., 1990, *The Arrow of Time*, (New York: Fawcett Columbine).

Coxeter, H. S. M, 1968 1999, *The Beauty of Geometry: Twelve Essays*, (New York: Dover Publications).

Crease, R. and Mann, C., 1986, *The Second Creation: Makers of the Revolution in Twentieth-Century Physics*, (New York: Macmillan).

Cropper, W., 2001, *Great Physicists*, (Oxford: Oxford University Press).

Crosby, Alfred, 1997 1998, *The Measure of Reality: Quantification and Western Science (1250-1600)*, (Cambridge, UK: Cambridge University Press).

Cuny, Hilaire, 1962 1965, *Albert Einstein: The Man and His Theories*, (New York: Paul S. Eriksson, Inc.).

D' Abro, A., 1927 1950, *The Evolution of Scientific Thought*, Second Edition, (New York: Dover Publications).

D' Inverno, Ray, 1992, *Approaches to Numerical Relativity*, (Cambridge: Cambridge University Press).

Darling, David, 1989, *Deep Time*, (New York: Delacorte Press, Bantom Doubleday).

Das, T. K., 1990, *The Time Dimension*, Bibliography, (New York: Praeger).

Dauber, P. and Muller, R., 1996, *The Three Big Bangs*, (New York: Addison-Wesley Publishing Company).

Davies, P. C. W. (Paul), 1977, *Space and Time in the Modern Universe*, (London: Cambridge University Press).

Davies, Paul, 1980, *Other Worlds*, (New York: Simon & Schuster).

Davies, Paul, 1981, *The Edge of Infinity*, (New York: Simon and Schuster).

Davies, P. C. W. (Paul) and Brown, J., Editors, 1988 1992, *Superstrings: A Theory of Everything*, (Cambridge: Cambridge University Press).

Davies, Paul, 1994, *The Last Three Minutes*, (New York: Basic Books).

Davies, Paul, 1995 1996, *About Time: Einstein's Unfinished Revolution*, (New York: Simon & Schuster).

Davies, Paul, 2002, *How to Build a Time Machine*, (New York: Penguin Books).

De Broglie, Louis, 1955, *Physics and Microphysics*, (New York: Pantheon Books).

De Broglie, Armand, Simon, et. Al., 1966 1979, *Einstein*, (New York: Peebles Press).

De Carle, D., 1965, *Horology*, (London: The English Universities Press).

De Caus, Salomon, 1624, *La Practique Et Demonstration Des Horloges Solaires*, (Paris: Chez Hyerosme Drouart).

De Vries, H., 1926, *Die Vierte Dimension*, (Berlin: Verlag und Druck von B. G. Teubner).

Delaney, William, 2004, *Discrete Event Physics: Space and Time*, (New York: I Universe).

Denbigh, Kenneth, 1981, *Three Concepts of Time*, (Berlin: Springer-Verlag).

Derbyshire, John, 2003, *Prime Obsession: Bernhard Riemann and the Greatest Unsolved Problem in Mathematics*, (Washington, D.C.: National Academies Press).

Descartes, Rene, 1677, *Collected Works of Descartes*, Two Volumes, (Amstelodami: Danielem Elsevirium).

Descartes, Rene, 1925 1954, *The Geometry of Rene Descartes: Translated by Smith and Latham*, (New York: Dover Publications).

Deutsch, David, 1997, *The Fabric of Reality*, (New York: The Penguin Press).

Devlin, Keith, 1998, *The Language of Math*, (New York: W. H. Freeman and Company).

Dewdney, A. K., 1984, *The Planiverse*, (New York: Simon and Schuster).

Digby, J., and Brier, B., Editors, 1985, *Permutations*, (New York: Quill).

Disalle, Robert, 2006, *Understanding Space-Time*, (Cambridge: Cambridge University Press).

Doehlemann, Karl, 1922 1924, *Projekitive Geometie in Synthetischer Behandlund...*, Fifth Edition, (Berlin: Walter De Gruyter).

Douglas, A. Vibert, 1957, *The Life of Arthur Stanley Eddington*, (New York: Thomas Nelson).

Du Sautoy, 2003, *The Music of the Primes*, (New York: Perennial).

Dugas, Rene, 1955 1988, *A History of Mechanics*, (New York: Dover Publications).

Dukas, H. and Hoffman, B., Editors, 1979, *Albert Einstein the Human Side: New Glimpses from His Archives*, (Princeton: Princeton University Press).

Dunne, J. W., 1927 1929, *An Experiment with Time*, (London: A & C Black, Ltd.).

Dunne, J. W., 1934, *The Serial Universe*, (London: Faber & Faber Ltd.).

Durell, Clement, 1926 1966, *Readable Relativity*, (London: G. Bell & Sons Ltd.).

Dusek, Val, 1999, *The Holistic Inspirations of Physics*, (New Brunswick: Rutgers University Press).

Dyckman, Dan, 1994, *Hidden Dimensions*, (London: Stanley Paul).

Dyson, Sir F. W., Eddington, A. S. and Davidson C., 1919 1921, *A Determination of the Deflection of Light by the Suns's Gravitational Field*, (Washington D.C.: The Smithsonian Institution).

Earman, J., 1995, *Bangs, Crunches, Whimpers, and Shrieks*, (Oxford: Oxford University Press).

Eddington, A. S., 1923 1937, *The Mathematical Theory of Relativity*, (London: Cambridge University Press).

Eddington, Sir Arthur, 1920 1959, *Space, Time & Gravitation*, (London: Cambridge University Press).

Eddington, Sir Arthur, 1933, *The Expanding Universe*, (London: Cambridge/ Macmillan).

Eddington, Sir Arthur S., 1920, *Physical Society of London: Report on the Relativity Theory of Gravitation*, (London: Fleetway Press).

Eddington, Sir Arthur, 1933 1974, *The Nature of the Physical World*, (Ann Arbor, MI: The University of Michigan Press).

Eddington, Sir Arthur Stanley, 1939 1940, *The Philosophy of Physical Science*, (New York: Macmillan, Cambridge University Press).

Edwards, H. M., 1974 2001, *Riemann's Zeta Function*, (New York: Dover Publications).

Edwards, Paul, Editor, 1964 1979, *Problems of Space and Time*, (New York: Macmillan Publishing).

Ehrlich, Robert, 1990, *Turning the World Inside Out*, (Princeton: Princeton University Press).

Einstein, Albert, 1905, "Zur Elektrodynamik bewegter Körper," *Annalen der Physik*, 17 (No. 10): 891-921.

Einstein, Albert, 1912 1996, 1912-1914 *Manuscript Facsimile on the Special Theory of Relativity*, (New York: George Braziller, Publishers).

Einstein, Albert, 1920, *Ather und Relativitatstheorie*, (Berlin: Julius Springer).

Einstein, 1920 1921, *Relativity: The Special and the General Theory*, (London: Methuen & Co., Ltd.).

Einstein, Albert, 1920 1961, *Relativity: The Special and the General Theory*, Fifteenth Edition, (New York: Crown Publishers).

Einstein, Albert, 1922 1983, *Sidelights on Relativity*, (New York: Dover Publications).

Einstein, Albert, 1922 1988, *The Meaning of Relativity*, Fifth Edition, (Princeton: Princeton University Press).

Einstein, Albert, 1927, *Zu Kaluzas Theorie des Zusammenhanges von Gravitation and Elektrizitat*, Offprint, (Berlin: Akademie Der Wissenschasten).

Einstein, Albert, 1928, *Riemann-geometrie Mit Aufrechterhaltug Des Begriffes Des Fernparallelismus*, (Berlin: Walter De Gruyter U. Co.).

Einstein, Albert, and Mayer, Walter, 1931, *Einheitliche Theorie von Gravitation and Elektrizitat / Einheitliche Theorie von Gravitation und...*, Offprint, (Berlin: Sitzungsberichten der Presussichen Akadimie).

Einstein, Albert, 1933 1934, *Essays in Science*, (New York: Philosophical Library).

Einstein, Albert, 1933 1949, *The World as I See It*, (New York: Philosophical Library).

Einstein, A. and Infeld, L., 1938, *The Evolution of Physics*, (New York: Simon and Schuster).

Einstein, Albert, 1950, *Theory of Relativity and Other Essays*, (New York: MJF Books).

Einstein, Albert, 1950, *Out of My Later Years*, (New York: Philosophical Library).

Einstein, Albert, 1952 1991, *Relativity*, (Chicago: Great Books).

Einstein, Albert, 1954, *Ideas and Opinions*, (New York: Crown Publishers).

Einstein, Albert, 1960, *Bibliographic Checklist and Index to the Published Writings of Albert Einstein*, (Paterson, NJ: Pageant Books, Inc.).

(Einstein), 2004, Special Einstein Issue, *Discover*, September.

Eisenhart, Luther Pfahler, 1926, *Riemannian Geometry*, (Princeton: Princeton University Press).

Ellis, George, 1988, *Flat and Curved Space-Times*, (Oxford: Oxford University Press).

Ellis, J. M. E., 1934, *Philosophical Studies*, (London: Edward Arnold & Co.).

Epstein, Lewis, 1981 1997, *Relativity Visualized*, (San Francisco: Insight Press).

Esposito, F. Paul and Witten, Louis, Editors, 1976 1977, *Asymptotic Structure of Space-Time*, (New York: Plem Press).

Euclid, 1566, *Analyseis Geometricae*, 1566, Original Work in Latin and Greek, First Six Books, (Strassburg: Dasypodius, Ed.).

Examined Life, The, Video Series, 2002, *Is Time Real? Does God Exist?*, (Pasadena, CA: Intelecom).

Euclid, 1533 1573, *Euclidis Elementorum, Libri X V, Graecee & Latinee*, (Corrected Reimpression of Cavellet's Edition).

Euclid, 300BC 1956, *The Thirteen Books of the Elements*, Volume 3, Books 10-13, (New York: Dover Publications).

Feather, Norman, 1959, *Introduction to the Physics of Mass Length and Time*, (London: Edinburgh University Press).

Felsager, Bjorn, 1981, *Geometry, Particles and Fields*, (Copenhagen: Odsense University Press).

Feltz, Michael, 1993, *Cosmo 101: The Four- Dimensional Universe*, (New York: Cosmos Publishing).

Ferguson, Kitty, 1991, *Black Holes in Spacetime*, (New York: Franklin Watts).

Ferguson, Kitty, 1999, *Measuring the Universe: Our Historic Quest to Chart the Horizons of Space and Time*, (New York: Walker and Company).

Ferris, Timothy, 1977, *The Red Limit*, (New York: William Morrow and Company, Inc.).

Ferris, Timothy, 1988, *Coming of Age in the Milky Way*, (New York: Doubleday Books).

Ferris, Timothy, 1997, *The Whole Shebang*, (New York: Simon & Schuster).

Ferris, Timothy, Editor, 1991, *The World Treasury of Physics, Astronomy, and Mathematics*, (Boston: Little Brown and Company).

Feuer, Lewis, 1974, *Einstein and the Generations of Science*, (New York: Basic Books).

Feynman, Michelle, 2005, *Perfectly Reasonable Deviations From the Beaten Track: The Letters of Richard P. Feynman*, (New York: Basic Books).

Feynmann, Richard, 1961, *The Feynman Lectures on Physics, Audio Recording, #20*, (New York: Pereus).

Feynman, Richard, 1963 1997, *Six Not-So-Easy Pieces*, (New York: Addison Wesley).

Feynman, Richard, 1965 1994, *The Character of Physical Law*, (New York: Random House).

Feynman, Richard, 1985, *Q E D: The Strange Theory of Light and Matter*, (Princeton: Princeton University Press).

Feynman, Richard, 1998, *The Meaning of It All*, (New York: Addison-Wesley Publishers).

Flegg, H. Graham, 1974 2001, *From Geometry to Topology*, (New York: Dover Publications).

Flew, Antony, Editor, 1956, *Essays in Conceptual Analysis*, (London: Macmillan & Co. Ltd.).

Fluckiger, Max, 1974, *Albert Einstein in Bern*, (Bern: Paul Haupt).

Fokker, Adriaan Daniel, 1922, *Het Relativiteitsbeginsel Voor Eenparige Translaties (1910-1912)*, (Leiden: Brill).

Fokker, Adriaan Daniel, 1929, *Relativeitstheorie*, (Groningen: P. Noordhoff).

Fokker, Adriaan Daniel, 1965, *Time and Space, Weight and Inertia*, (Oxford: Pergamon Press).

Folsing, 1993 1997, *Albert Einstein*, (New York: Viking).

Ford, Henry, Trade School, *An Introduction to the Fourth Dimension,* (Dearborn, MI: Henry Ford Trade School).

Fraknoi, Andrew, Editor, 2000, *Cosmos in the Classroom 2000,* (San Francisco: The Astronomical Society of the Pacific).

Frank, Philipp, 1947 1953, *Einstein His Life and Times,* (New York: Plenum Publishing Corporation).

Frank, Philipp, 1957 1974, *Philosophy of Science,* (New York: Greenwood Press Reprint).

Frank, Philipp, 1943, *Relativity and its Astronomical Implications,* Booklet, (Harvard: Sky Publishing Corporation).

Fraser, J. T. Editor, 1966, *The Voices of Time,* (New York: George Braziller).

Fraser, J. T., 1982, *The Genesis and Evolution of Time,* (Amherst: University of Massachusetts Press).

French, A. P., 1966 1968, *Special Relativity,* (New York: M.I.T., Norton & Company).

French, A. P., Editor, 1979, *Einstein: A Centenary Volume,* (Cambridge: Harvard University Press).

Friedman, Michael, 1983, *Foundations of Space-timeTheory,* (Princeton: Princeton University Press).

Friedman, William, 1990, *About Time: Inventing the Fourth Dimension,* (Cambridge: The M.I.T. Press).

Fuller, Buckminster, 1975, *Synergetics: Explorations in the Geometry of Thinking,* (New York: Macmillan Publishing Co., Inc.).

Fuller, Buckminster, 1975 1982, *Tetrascroll: A Cosmic Fairy Tale,* (New York: St. Martin's Press).

Galison, Peter, 2003, *Einstein's Clocks, Poincare's Maps,* (New York: Norton & Co).

Gamov, George, 1940 1942, *Mr. Tompkins in Wonderland,* (New York: Macmillan).

Gamow, George, 1940 1965, *Mr. Tompkins in Paperback,* (Cambridge: Cambridge University Press).

Gamow, George, 1947 1948, *One Two Three...Infinity,* (New York: Viking Press).

Gamov, George, 1961 1988, *Great Physicists From Galileo to Einstein,* (New York: Dover Publications).

Gamov, George, 1962, *Gravity*, (New York: Doubleday & Company).

Gamow, George, 1966 1985, *Thirty Years that Shook Physics*, (New York: Dover Publications).

Gardenfors, Peter, 2000, *Conceptual Spaces*, (Cambridge, MA: The M.I.T. Press).

Gardner, Martin, 1962, *Relativity for the Million*, (New York: Macmillan).

Gardner, Martin, 1962 1996, *Relativity Simply Explained*, (New York: Dover Publications).

Gardner, Martin, 1962 1976, *The Relativity Explosion*, (New York: Vintage Books).

Gardner, Martin, 1964 1979, *The Ambidextrous Universe*, (New York: Charles Scribner's Sons).

Gardner, Martin, 1996, *The Night Is Large*, (New York: St. Martin's Press).

Gefter, Amanda, 2005, Putting Einstein to the Test, *Sky & Telescope*, July, 32-40.

Gellner, Ernest, 1964, *Thought and Change*, (London: University of Chicago Press).

Geroch, Robert, 1978 1981, *General Relativity from A to B*, (Chicago: University of Chicago Press).

Geroch, Robert, 1985, *Mathematical Physics*, (Chicago: University of Chicago Press).

Giancoli, Douglas, 1984 2000, *Physics for Scientists and Engineers*, Third Edition, (New York: Prentice Hall).

Gibbons, Shellard, and Rankin, Editors, 2003, *The Future of Theoretical Physics and Cosmology: Celebrating Stephen Hawking's 60th Birthday*, (Cambridge: Cambridge University Press).

Gibilisco, Stan, 1983, *Understanding Einstein's Theories of Relativity*, (Blue Ridge Summit: Tab Books, Inc.).

Gilmore, Robert, 1995, *Alice in Quantumland*, (New York: Springer-Verlag).

Goldberg, Stanley, 1984, *Understanding Relativity: Origin and Impact of a Scientific Revolution*, (Boston, Stuttgart: Birkhauser Boston, Inc.).

Goldfarb, Warren D., Editor, 1971, *Jacques Herbrand Logical Writings of 1928*, (Cambridge, M A: Harvard University Press).

Goldsmith, Donald, 1995, *Einstein's Greatest Blunder?*, (Cambridge, MA: Harvard University Press).

Goldsmith, Donald, 1997, *The Ultimate Einstein*, (New York: Byron Preiss Multimedia Books).

Goldstein and Goldstein, 1978, *How We Know*, (New York: Plenum Paperback).

Gott, J. Richard, 2001, *Time Travel in Einstein's Universe*, (New York: Houghton Mifflin Company).

Gray, J., Editor, 1999, *The Symbolic Universe*, (Oxford: Oxford University Press).

Gray, Jeremy, 2004, Janos Bolyai, *Non-Euclidean Geometry, and the Nature of Space*, (Cambridge, MA: Burndy Library Publications).

Greenberg, Marvin Jay, 1993, *Euclidean and Non-Euclidean Geometries, Development and History*, (New York: W. H. Freeman).

Greene, Brian, 2004, *The Fabric of the Cosmos*, (New York: Alfred A. Knopf).

Gregory, Bruce, 1987 1990, *Inventing Reality*, (New York: John Wiley & Sons).

Gribbin, John, 2001, *Hyperspace Our Final Frontier*, (New York: DK Publishing, Inc.).

Gribbin, John, 1984 1988, *In Search of Schrodinger's Cat*, (New York: Bantam Books).

Gribbin, John, 1995, *Schroedinger's Kittens and the Search for Reality*, (Boston: Little Brown and Company).

Gribbin, John, 1992, *Unveiling the Edge of Time: Black Holes, White Holes, Wormholes*, (New York: Harmony Books/ Crown Publishers).

Gribbin, Mary and Gribbin, John, 1994, *Time & Space*, (London: Eyewitness, D. K. Dorling Kindersley).

Griffin, David, Editor, 1986, *Physics and the Ultimate Significance of Time*, (Albany, N Y: State University of New York Press).

Griffiths, David J., 1981, *Introduction to Electrodynamics*, (New York: Prentice Hall).

Griffiths, Jay, 1999, *A Sideways Look at Time*, (New York: Jeremy P. Tarcher, Putnam).

Grunbaum, Adolf, 1963, *Philosophical Problems of Space and Time*, (New York: Alfred A. Knopf).

Grunbaum, Adolf, 1968, Reprint, *Reply to Hilary Putnam's "An Examination of Grunbaum's Philosophy of Geometry"*, (Dordrecht, Holland: D. Reidel Publishing Company).

Guggenheimer, Samuel, 1925, *The Einstein Theory Explained and Analyzed*, (New York: Macmillan Company).

Guillen, Michael, 1995, *Five Equations That Changed the World*, (New York: Hyperion).

Guthrie, W. K. C., 1967, *A History of Greek Philosophy: Volume 1, The Earlier Presocratics and the Pythagoreans*, (London: Cambridge University Press).

Hamlyn, D. W., 1984, *Metaphysics*, (Cambridge, England: Cambridge University Press).

Hancock, Harris, 1939 1964, *Development of the Minkowski Geometry of Numbers*, Two Volumes, (New York: Dover Publications).

Hanlon, A. C., 1977 1985, *Introduction to the Fourth Dimension*, (India: Theosophical Publishing House).

Harrison, Edward, 1985, *Masks of the Universe*, (New York: Macmillan).

Harrow, Benjamin, 1920, *From Newton to Einstein*, (New York: Van Nostrand).

Hatton, J. and Plouffe, P., 1997, *Science and its Ways of Knowing*, (Upper Saddle River: Prentice Hall).

Hausner, Melvin, 1965 1998, *A Vector Space Approach to Geometry*, (New York: Dover Publications).

Hawking, S. W. & Ellis, G. F. R., 1973 1980, *The Large Scale Structure of Space-Time*, (Cambridge: Cambridge University Press).

Hawking, Stephen, 1973 1995, *Large Scale Structure of Space-Time*, (Cambridge: Cambridge University Press).

Hawking, Stephen & Israel, W. & W. Editors, 1979, *General Relativity: An Einstein Centenary Survey*, (Cambridge, M A: Cambridge University Press).

Hawking, Stephen, 1988, *A Brief History of Time*, (New York: Bantam Books).

Hawking, Stephen, 1992, *A Brief History of Time: A Reader's Companion*, (New York: Bantam Books).

Hawking, Stephen, 1993, *Black Holes & Baby Universes and Other Essays*, (New York: Bantam Books).

Hawking, Stephen, 1996, *Nature of Space and Time*, (Princeton: Princeton University Press).

Hawking, Stephen, 2001, *The Universe in a Nutshell*, (New York: Bantam Books).

Hawking, S., Thorne, K. et. Al., 2002, *The Future of Spacetime*, (New York: W. W. Norton and Company).

Heiserman, David, 1983, *Experiments in Four Dimensions*, (New York: Tab Books, Inc.).

Held, A., Editor, 1980, *General Relativity and Gravitation. One Hundred Years After the Birth of Einstein*, Volume 1, (New York: Plenum Press).

Held, A., Editor, 1980, *General Relativity and Gravitation. One Hundred Years After the Birth of Einstein*, Volume 2, (New York: Plenum Press).

Helmholtz, Hermann Ludwig Ferdinand Von, 1885 1892, *Wessenschaftliche Abhandlungen*, (Leipzig: Johann Ambrosius Barth).

Henderson, Linda Dalrymple, 1983, *The Fourth Dimension and Non-Euclidean Geometry in Modern Art*, (Princeton: Princeton University Press).

Herbert, Nick, 1988, *Faster than Light*, (New York: Penguin Books).

Herneck, Friedrich, 1982, *Albert Einstein*, (Leipzig, Germany: B G Teubner).

Highfield, R. & Carter, P., 1993, *The Private Lives of Albert Einstein*, (New York: St. Martins Press).

Hilbert, David, 1902 1938, *The Foundations of Geometry*, (Chicago: Open Court Publishing).

Hilbert, David & Ackermann, W., 1928 1946, *Grundzuge Der Theoretischen/ die Grundlehren Der Mathematischen Wissenschaften*, (New York: Dover Publications).

Hilbert, David & Cohn-Vossen, S., 1952 1999, *Geometry and the Imagination*, (Providence, RI: American Mathematical Society).

Hilbert, David, 1993, *Theory of Algebraic Invariants*, (Cambridge: Cambridge Mathematical Library).

Hinton, C. H., 1904, *The Fourth Dimension*, (London: Swan Sonnenschein Ltd.).

Hinton, C. H., 1904, The Fourth Dimension, *Harper's Monthly Magazine*, July, 229-233.

Hinton, C. H., 1980 1993, *The Fourth Dimension*, (Ayer Company Publishers1980/Mokelumne Hill Press 1993).

Hodson, G. and Horne, A., 1933, *Some Experiments in Four-Dimensional Vision*, Facsimile, (Plymouth, England: The Mayflower Press).

Hoffman, Banesh, 1983, *Relativity and its Roots*, (New York: Scientific American Books).

Hoffmann, B. and Dukas, H., 1972, *Albert Einstein: Creator and Rebel*, (New York: New American Library).

Holland, Charles Hepworth, 1999, *The Idea of Time*, (New York: John Wiley & Sons).

Holton, G. and Elkana, Y., Editors, 1982, *Albert Einstein: Historical and Cultural Perspectives*, (Princeton: Princeton University Press).

Holton, Gerald & Elkana, Yehuda, Editors, 1984, *Albert Einstein: Historical and Cultural Perspectives*, (Princeton: Princeton University Press).

Holton, Gerald, 1995 1996, *Einstein, History, and Other Passions*, (New York: Addison- Wesley Publishing Company).

Holton, Jean Laity, 1971, *Geometry: A New Way of Looking at Space*, (New York: Weybright and Talley).

Horgan, John, 1996 1997, *The End of Science*, (New York: Bantam Doubleday Dell).

Hudgins, William, 1920, *An Introduction to Einstein's Theory of Relativity*, Booklet, (Girard, K S: Little Blue Books).

Hudgings, William, 1921, *Introduction to Einstein and Universal Relativity*, (New York: Arrow Book Company).

Hupfeld, Herman, 1931 1942, *As Time Goes By*, Casablanca Sheet Music, (New York: Harms Inc, Warner Brothers).

Hurley, Donal & Vandyck, Michael, 2000, *Geometry, Spinors and Applications*, (London: Springer- Praxis).

Husserl, Edmund, 1907 1973, *The Idea of Phenomenology*, (The Hague: Martinus Nijhoff).

Hutchins, Robert, Editor, 1934 1952, *Great Books of the Western World*, Volume #34, Newton and Huygens, (Chicago: Encyclopedia Britannica, Inc).

Indiana University Debate, Sigma Xi, 1927, *A Debate on the Theory of Relativity*, (Chicago: Open Court Publishing).

Infield, Leopold, 1950, *Albert Einstein: His Work and its Influence on Our World*, (New York: Charles Scribner's Sons).

Infield, Leopold, 1962, *Recent Developments in General Relativity - Essays Dedicated to Infield on His 60th Birthday*, (Warszawa, Poland: The Macmillan Company).

Ivins, Jr, William M, 1946 1964, *Art & Geometry*, (New York: Dover Publications).

Jammer, Max, 1954 1957, *Concepts of Space: The History of the Theories of Space in Physics*, (New York: Harvard University Press).

Jammer, Max, 1954 1993, *Concepts of Space*, Third Edition, (New York: Dover Publications).

Jammer, Max, 1957 1999, *Concepts of Force*, (New York: Dover Publications).

Jammer, Max, 1961 1997, *Concepts of Mass in Classical and Modern Physics*, (New York: Dover Publications).

Jammer, Max, 2000, *Concepts of Mass in Contemporary Physics and Philosophy*, (Princeton: Princeton University Press).

Jammer, Max, 2006, *Concepts of Simultaneity From Antiquity to Einstein, and Beyond*, (Baltimore: The Johns Hopkins University Press).

Jeans, Sir James, 1933 1953, *The New Background of Science*, (London: Cambridge University Press).

Jeans, Sir. James, 1934, *Through Space and Time*, (New York: Macmillan Company).

Jeans, Sir James, 1942, *Physics & Philosophy*, (Cambridge: Cambridge University Press).

Jespersen, J. and Fitz-Randolph, J., 1977 1999, *From Sundials to Atomic Clocks*, Second Edition, (New York: Dover Publications).

Johnson, M., 1947, *Time, Knowledge, and the Nebulae*, (New York: Dover Publications).

Johnson, Timothy, 1967 1968, *River of Time*, (New York: Coward- McCann, Inc.).

Katz, Robert, 1964, *Special Theory of Relativity*, (Princeton: D. Van Nostrand Company).

Kaku, Michio, 1987 1995, *Beyond Einstein*, (New York: Doubleday Books).

Kaku, Michio, 1994, *Hyperspace*, (New York: Doublebay).

Kaku, Michio, 1997, *Visions*, (New York: Doubleday Books).

Kaku, Michio 2004, Einstein in a Nutshell, Special Einstein Issue, *Discover*, September, 17-24.

Kaku, Michio, 2005, *Parallel Worlds*, (New York: Doubleday).

Kaufmann, William, 1973, *Relativity & Cosmology*, (New York: Harper & Row).

Kaufmann, William, 1979 1980, *Black Holes and Warped Spacetime*, (New York: Bantam Books).

Kelvin, Lord, and Guthrie, Peter, 1872 1894, *Elements of Natural Philosophy*, (Cambridge: University Press).

Kenyon, I. R., 1990 1996, *General Relativity*, (Oxford: Oxford University Press).

Kinnon, Colette, et. Al. Editors, 1981, *The Impact of Modern Scientific Ideas on Society in Commemoration of Einstein*, (Dordrecht, Holland: Reidel).

Klein, Felix, 1893 1963, *On Riemann's Algebraic Functions and Their Integrals, Supplement*, (New York: Dover Publications).

Klein, Christian Felix, 1911 1914, *Elementarmathematik Vom Hoheren Standpunkte Aus.*, Volumes 1 and 2, (Leipzig: B. G. Teubner).

Klein, Christian Felix, 1921 1922, *Gesammelte Mathematische Abhandlungen...*, (Berlin: Julius Springer).

Klein, Christian Felix, 1926, *Vorlesungen Uber Die Entwichlung Der Mathematik Im 19. Jahrhundert*, (Berlin: Julius Springer).

Klein, Christian Felix, 1928, *Vorlesungen Uber Nicht-euklidische Geometrie*, (Berlin: Verlag Springer).

Klein, Christian Felix, 1979, *Vorlesungen Uber Die Entwicklung Der Mathematik Im 19 Jahrhundert*, Volumes 1 and 2, Reprint, (Berlin: Springer-Verlag).

Klein, Etienne, 1988 1996, *Conversations with the Sphinx*, (London: Souvenir Press).

Kline, Morris, 1953, *Mathematics in Western Culture*, (New York: Oxford University Press).

Knopp, Konrad, 1948 2000, *Problem Book in the Theory of Functions*, (New York: Dover Publications).

Kopff, A., 1923, *The Mathematical Theory of Relativity*, (London: Methuen & Co. Ltd.).

Koslow, Arnold, Editor, 1967, *The Changeless Order: The Physics of Space and Time*, (New York: George Braziller, Inc.).

Krause, Eugene, 1975 1986, *Taxicab Geometry*, (New York: Dover Publication).

Krauss, Lawrence M., 2005, *Hiding in the Mirror*, (New York: Penguin Group).

Lanczos, Cornelius, 1965, *Albert Einstein and the Cosmic World Order*, (New York: John Wiley & Sons, Inc.).

Lanczos, Cornelius, 1974, *The Einstein Decade (1905-1915)*, (New York: Academic Press).

Landes, David, 1983, *Revolution in Time*, (Cambridge, MA: Harvard University Press).

Lang, K. and Gingerich, O., Editors, 1979, *A Source Book in Astronomy and Astrophysics, 1900-1975*, (Cambridge, MA: Harvard University Press).

Lange, Marc, 2000, *Natural Laws in Scientific Practice*, (Oxford: Oxford University Press).

Langone, John, 2002, *The Mystery of Time: Humanity's Quest for Order and Measure*, (New York: National Geographic).

Laudan, Larry, 1990, *Science and Relativism*, (Chicago: University of Chicago Press).

Laughlin, Robert B., 2005, *A Different Universe: Reinventing Physics From the Bottom Down*, (New York: Basic Books).

Lazerowitz, M., 1955, *The Structure of Metaphysics*, (London: International Library of Psychology, Philosophy and Scientific Method).

Le Poidevin, Robin, 2003, *Travels in Four Dimensions*, (Oxford: Oxford University Press).

Lederman, Leon, 1993, *The God Particle*, (New York: Bantam Doubleday Dell Publishing).

Lemonick, Michael, 1993 1995, *The Light at the Edge of the Universe*, (Princeton: Princeton University Press).

Lerner, Aaron, 1973, *Einstein & Newton*, (Minneapolis, MN: Lerner Publications Company).

Leshan, L. and Margenau, H., 1982, *Einstein's Space & Van Gogh's Sky*, (New York: Macmillan Publishing Company).

Levenson, Thomas, 2003, *Einstein in Berlin*, (New York: Bantam Books).

Levine, Robert, 1997, *A Geography of Time*, (New York: Basic Books).

Levinger, Elma Ehrlich, 1949, *Albert Einstein*, (New York: Julian Messner, Inc.).

Lieber, Lillian R., 1936 1945, *The Einstein Theory of Relativity*, (New York: Rinehart & Co, Inc.).

Lightman & Brawer, 1990, *The Lives and Worlds of Modern Cosmologists*, (Cambridge: Harvard University Press).

Lightman, Alan, 1990 1991, *Ancient Light*, (Cambridge: Harvard University Press).

Lindgren, C. and Slaby, S., Editors, 1968, *Four-Dimensional Descriptive Geometry*, (New York: Mcgraw-Hill).

Lines, Malcolm E., 1994, *On the Shoulders of Giants*, (Bristol: Institute of Physics Publishing).

Lineweaver, C. H. and Davis, T. M., 2005, Misconceptions about the Big Bang, *Scientific American*, March, 36-45.

Lobacheveski, Nicholas, 1840 1914, *Geometric Researches on the Theory of Parallels*, (Chicago: Open Court Publishing).

Lodge, Sir Oliver, 1909, *The Ether of Space*, (New York: Harper & Bros).

Lorentz, H. A., 1909, *The Theory of Electrons and its Applications to the Phenomena of Light and Radiant Heat*, (Leipzig: Teubner).

Lorentz, H. A., 1909, *Theory of Electrons*, (New York: Columbia University Press).

Lorentz, H. A., 1919 1920, *The Einstein Theory of Relativity*, (New York: Brentano's).

Lorentz, H. A., 1920, *Aethertheorieen en Aethermodellen* (1901-1902), (Leiden, Germany: Brill).

Lorentz, Einstein, Weyl, Minkowski et. Al., 1923 1952, *The Principle of Relativity*, (New York: Dover Publications).

Lucas, J. R. and Hodgson, P. E., 1990, *Spacetime & Electromagnetism*, (Oxford: Oxford University Press).

Ludvigsen, M., 1999, *General Relativity: A Geometric Approach*, (Cambridge: Cambridge University Press).

Luminet, Jean- Pierre, 1987 1992, *Black Holes*, (Cambridge: Cambridge University Press).

Macey, S., Editor, 1994, *Encyclopedia of Time*, (New York: Garland Reference Library).

Mach, Ernst, 1883 1921, *Die Mechanik in Ihrer Entwicklung*, (Leipzig: Brockhaus).

Mach, Ernst, 1906 2004, *Space and Geometry*, (New York: Dover Publications).

Macvey, John, 1990, *Time Travel: A Guide to Journeys in the Fourth Dimension*, (Chelsea, MI: Scarborough House).

Magie, W., Editor, 1935 1969, *A Source Book in Physics*, (Cambridge, MA: Harvard University Press).

Magueijo, Joao, 2003 2004, *Faster than the Speed of Light*, (New York: Penguin Books).

Mandelker, Jakob, 1966, *Relativity and the New Energy Mechanics*, (New York: Philosophical Library).

Manning, H P Editor, 1910 1960, *The Fourth Dimension Simply Explained*, (New York: Dover Publications).

Manning, Henry Parker, 1914 1956, *Geometry of Four Dimensions*, (New York: Dover Publications).

Manning, Henry Parker, 1914, *Geometry of Four Dimensions*, (New York: The Macmillan Company).

Manoff, Sawa, 2002, *Geometry and Mechanics in Different Models of Space-Time*, (New York: Nova Science Publishers, Inc.).

Marder, L., 1971 1974, *Time and the Space-Traveler*, (New York: University of Pennsylvania Press).

Marianoff, D. and Wayne, P., 1944, *Einstein*, (New York: Doubleday, Doran and Co.).

Marshall, I. and Zohar, D., 1997, *Who's Afraid of Schrodinger's Cat?*, (New York: William Morrow).

Martin, Herbert, 1947, *The Inquiring Mind: Introduction to Philosophic Studies*, (New York: Barnes & Nobles).

Maxwell, James Clerk, 1881, *A Treatise on Electricity and Magnetism*, (Oxford: Oxford).

Mazur, Barry, 2002 2004, *Imaging Numbers: Particularly the Square Root of -15*, (New York: St. Martins Press).

McCleary, John, 1994, *Geometry from a Differentiable Viewpoint*, (Cambridge: Cambridge University Press).

McCready, Stuart, Editor, 2001, *Time: The Discovery of Time*, (Naperville, IL: Sourcebooks, Inc.).

McGilvary, E. B., 1956, *Toward a Perspective Realism*, (Chicago: Open Court Publishing).

McGrath, J. and Kelly, J., 1986, *Time & Human Interaction: Toward a Social Psychology of Time*, (New York: The Guilford Press).

McInerney, Peter, 1991, *Time and Experience*, (Philadelphia: Temple University Press).

McLeish, John, 1991, *Number: The History of Numbers and How They Shape Our Lives*, (New York: Fawcett Columbine).

McTaggart, J. M. E., 1927 1968, *The Nature of Existence*, Volume 2, (London: Cambridge University Press).

Meerloo, Dr. J. A. M., 1970, *Along the Fourth Dimension: Man's Sense of Time and History*, (New York: John Day Co.).

Meursius, Johannes, 1631, *Denarivs Pythagoricvs, Pythagorus and Music of the Spheres*, (Leyden: Lugduni Batavorum).

Michaelmore, Peter, 1962, *Einstein: Profile of the Man*, (New York: Dodd, Mead & Company).

Michalos, Alex, Editor, 1974, *Philosophical Problems of Science & Technology*, (Boston: Allyn & Bacon, Inc.).

Mill, John Stuart, 1869, *Analysis of the Phenomena of the Human Mind*, Two Volumes, (London: Longmans, et. Al.).

Miller, Arthur, 1981, *Albert Einstein's Special Theory of Relativity: Emergence (1905) and Early Interpretation (1905-1911)*, (London: Addison Wesley).

Miller, Arthur, 1977 1981, *Albert Einstein's Special Theory of Relativity*, (New York: Springer).

Miller, Arthur I., 1984, *Imagery in Scientific Thought*, (Boston: Birkhauser).

Milne, E. A., Editor, 1935, *Relativity Gravitation and World-Structure*, (Oxford: Oxford University Press).

Minkowski, Hermann, 1899, *Ein Kriterium Fur die Algebraischen Zahlen*, Offprint, (Gottingen: Teubner).

Minkowski, Hermann, 1899, *Ueber die Annaherung an Eine Reelle Grosse Durch Rationale Zahlen 1899*, Offprint, (Leipzig: Mathematische Annalen).

Minkowski, Hermann, 1900, *Quelques Nouveaux Theoremes Sur L'approx - imation Des Quantites a L'aide de Numbres Rationnels*, (Paris: Gauthier-Villars).

Minkowski, Hermann, 1904 1905, *Zur Geometrie der Zahlen 1905*, Offprint, (Leipzig: Teubner).

Minkowski, Hermann, 1909, *Raum et Zeit* (Original Publication), (Leipzig und Berlin: Druck und Verlag Von B. G. Teubner).

Misner, C., Thorne, K., and Wheeler, J., 1970 1973, *Gravitation*, (New York: Freeman & Co.).

Miyazaki, Koji, 1983 1986, *An Adventure in Multidimensional Space*, (New York: John Wiley & Sons).

Mlodinow, Leonard, 2001, *Euclid's Window*, (New York: Simon and Schuster).

Mook, D. and Vargish, T., 1987, *Inside Relativity*, (Princeton: Princeton University Press).

Moore, George Edward, 1953, *Some Main Problems of Philosophy*, (New York: The Macmillan Company).

Moore, Pete, 2002, *E=mc²: The Great Ideas that Shaped Our World*, (Toronto: Friedman/ Fairfax).

Moore, Thomas, 2003, *Six Ideas That Shaped Physics*, Second Edition, (New York: McGraw-Hill).

Morgan, Frank, 1998, *Riemannian Geometry, A Beginner's Guide*, Second Edition, (Wellesley, MA: A. K. Peters).

Morris, Richard, 1993, *Cosmic Questions*, (New York: John Wiley & Sons).

Morris, Richard, 1985, *Time's Arrows: Scientific Attitudes Toward Time*, (New York: Simon & Schuster).

Mundle, C. W. K., 1954, "Mr. Dobbs' Two-Dimensional Theory of Time," *The British Journal for the Philosophy of Science*, (Notes and Comments), 4 (February): 331-337.

Munitz, Milton, 1986, *Cosmic Understanding*, (Princeton: Princeton University Press).

Munitz, Milton, Editor, 1957, *Theories of the Universe: Babylonian Myth to Modern Science*, (Glencoe, IL: The Free Press).

Murchie, Guy, 1961, *Music of the Spheres*, (Cambridge, MA: Riverside Press Cambridge).

Naber, Gregory, 1992 2003, *The Geometry of Minkowski Spacetime*, (New York: Dover Publications).

Nagel, Ernest, 1961, *The Structure of Science - Problems in the Logic of Scientific Explanation*, (New York: Harcourt, Brace & World, Inc.).

Nahin, Paul J., 1993, *Time Machines: Time Travel in Physics, Metaphysics, and Science Fiction*, (New York).

Nathan, O. and Norden, H., Editors, 1960, *Einstein on Peace*, (New York: Avenel Books).

Newman, James R., Editor, 1956, *The World of Mathematics*, Four Volumes, (New York: Simon and Schuster).

Nicod, Jean, 1930 1950, *Foundations of Geometry & Induction*, (London: Routledge & Kegan Paul Ltd.).

Novikov, I., 1998 2001, *The River of Time*, (Cambridge, England: Cambridge University Press/ Canto).

O'Rahilly, Alfred, 1938 1965, *Electromagnetic Theory: A Critical Examination*, Volume 2, (New York: Dover Publications).

Orstein, Robert, 1969 1975, *On the Experience of Time*, (New York: Penguin Books).

Osserman, Robert, 1995, *Poetry of the Universe*, (New York: Anchor Books, Doubleday).

Ouspensky, P. D., 1920 1966, *Tertium Organum*, (New York: Alfred Knoff).

Ouspensky, P. D., 1931 1997, *A New Model of the Universe*, (New York: Dover Publications).

P B S Home Video, 1991, *A. Einstein: How I See the World*, Video Tape, (Los Angeles: Pacific Arts Video).

Pagels, Heinz R., 1985, *Perfect Symmetry*, (New York: Simon and Schuster).

Pais, Abraham, 1982, *Subtle Is the Lord...: The Science and the Life of Albert Einstein*, (New York: Oxford University Press).

Pais, Abraham, 1991, *Niels Bohr's Times*, (Oxford: Clarendon Press, Oxford University Press).

Pais, Abraham, 1994, *Einstein Lived Here*, (Oxford: Oxford University Press).

Park, David, 1997, *The Fire Within the Eye*, (Princeton: Princeton University Press).

Parker, B., 2000, *Einstein's Brainchild*, (Amherst, N Y: Prometheus Books).

Paterniti, Michael, 2000, *Driving Mr. Albert*, (New York: Random House, Inc.).

Patrides, C. A., Editor, 1976, *Aspects of Time*, (Manchester, England: Manchester University Press).

Pauli, W., 1921 1958, *Theory of Relativity*, (New York: Dover Publications).

Pawlicki, T B, 1984, *How You Can Explore Higher Dimensions of Space and Time*, (New York: Prentice-Hall, Inc.).

Peat, F. David, 1990, *Einstein's Moon: Bell's Theorem and the Curious Quest for Quantum Reality*, (Chicago: Contemporary Books).

Peebles, P. J. E., 1993, *Principles of Physical Cosmology*, (Princeton: Princeton University Press).

Penrose & Rindler, 1984 1986, *Spinors and Space-Time*, Volume 1, (Cambridge, England: Cambridge University Press).

Penrose & Rindler, 1986, *Spinors and Space-Time*, Volume 2, (Cambridge, England: Cambridge University Press).

Penrose, R. & Isham, C. J., Editors, 1986, *Quantum Concepts in Space and Time*, (Oxford: Clarendon Press Oxford).

Penrose, Roger, 1989, *The Emperor's New Mind*, (New York: Oxford University Press).

Penrose, Roger, 1994, *Shadows of the Mind*, (Oxford: Oxford University Press).

Penrose, Roger, 1995 1999, *The Large, the Small, and the Human Mind*, (Cambridge: Cambridge University Press).

Penrose, R., Huggett, et. Al. Editors, 1998, *The Geometric Universe: Science, Geometry, and the Work of Roger Penrose*, (Oxford: Oxford University Press).

Penrose, Roger, 2005, *The Road to Reality*, (New York: Alfred A. Knopf).

Perkowitz, Sidney, 1996, *Empire of Light*, (New York: Henry Holt and Company).

Perlmutter, Arnold & Linda, Scott, Linda, Editors, 1979, *On the Path of Albert Einstein*, (New York: Plenum Press).

Petersen, Peter, 1998, *Riemannian Geometry*, (London: Springer).

Pickover, Clifford, 1996, *Black Holes A Travel Guide*, (New York: John Wiley & Sons, Inc.).

Pickover, C., 1998 1999, *Time: A Traveler's Guide*, (Oxford: Oxford University Press).

Pickover, Clifford, 1999, *Surfing Through Hyperspace: Understanding Higher Universes in Six Easy Lessons*, (London: Oxford University Press).

Plank, Max, 1908, *Zur Dynamik Bewegter Systeme*, Offprint, (Leipzig: Annalen Der Physik).

Plank, Dr. Max, 1910, *Theoretishche Physik*, (Leipzig: Verlag Von S. Hirzel).

Poincare, H., 1913 1946, *The Foundations of Science*, (Lancaster, PA: The Science Press).

Poincare, Henri, 1913 1952, *Science and Method*, (New York: Dover Publications).

Poor, Charles, 1922, *Gravitation Versus Relativity*, (New York: G. P. Putnam's Sons).

Poulet, Georges, 1956 1959, *Studies in Human Time*, (New York: Harper Torchbooks).

Powell, Corey S., 2002, *God in the Equation*, (New York: The Free Press/ Simon & Schuster).

Price, Huw, 1996, *Time's Arrow & Archimedes' Point*, (Oxford: Oxford University Press).

Priestley, J. B., 1964, *Man and Time*, (London: Bloomsbury Books).

Putnam, Hilary, 1975, *Mathematics Matter and Method, Philosophical Papers*, Volume I, (London: Cambridge University Press).

Putnam, Hilary, 1975, *Mind Language and Reality*, Philosophical Papers, Volume II, (London: Cambridge University Press).

Pyenson, 1985, *The Young Einstein*, (Bristol:).

Pythagoras, Gathered by Iambichus of Chalcis, 1598, De Vita Pythagorica, (Heidelberg: Bibliopolio Commeliniano).

Raine, D. and Heller, M., 1981, *The Science of Space-Time*, (Tucson, AZ: Pachart Publishing House).

Randall, Lisa, 2005, *Warped Passages: Unraveling the Mysteries of the Universe's Hidden Dimensions*, (New York: Harper Collins Publishers).

Ray, Christopher, 1987, *The Evolution of Relativity*, (Bristol, England: Adam Hilger).

Ray, M., 1965, *Theory of Relativity: Special and General*, (Delhi, India: S. Chand & Co.).

Razelos, Panagiotis, Editor, 1982, *The Einstein Centennial*, (New York: College of Staten Island, City University of New York).

Rees, Sir. Martin, 1995 2000, *New Perspectives in Astrophysical Cosmology*, (Cambridge, England: Cambridge University Press).

Rees, Sir. Martin, 1999 2000, *Just Six Numbers*, (New York: Basic Books).

Rees, Sir. Martin, 2001, *Our Cosmic Habitat*, (Princeton: Princeton and Oxford).

Regis, Ed, 1987, *Who Got Einstein's Office*, (New York: Addison-Wesley Publishing Company).

Reichenbach, H., 1956 1999, *The Direction of Time*, (New York: Dover Publications).

Reichenbach, Hans, 1952, *From Copernicus to Einstein*, (New York: Philosophical Library).

Reichenbach, Hans, 1927 1958, *Philosophy of Space & Time*, (New York: Dover Publications).

Reichinstein, D., 1932 1934, *Albert Einstein*, (London: Edward Goldston Ltd.).

Reid, Constance, 1996, *Hilbert: David Hilbert Biography*, (New York: Copernicus/Springer-Verlag).

Resnick, R. and Halliday, D., 1972 1992, *Basic Concepts in Relativity*, (New York: Macmillan Publishing Company).

Rice, James, 1929, *Relativity: An Exposition without Mathematics*, (New York: Jonathan Cape & Harrison Smith).

Ridley, B., 1976 2000, *Time, Space and Things*, Third Edition, (Cambridge, England: Cambridge University Press/Canto).

Riemann, Bernhard, 1892, *Collected Works of Bernard Riemann*, Second Edition, (Leipzig: B. G. Teubner).

Riemann, Bernhard, 1892 1953, *Collected Works of Bernard Riemann*, Second Edition, Facsimile, (New York: Dover Publications).

Rifkin, Jeremy, 1987, *Time Wars*, (New York: Henry Holt and Company).

Robbin, Tony, 1992, *Fourfield: Computers, Art & the 4th Dimension*, (Boston: Little Brown & Co.).

Robert Kaplan and Ellen Kaplan, 2003, *The Art of the Infinite: The Pleasures of Mathematics*, (Oxford: Oxford University Press).

Robinson, Enders, 1990, *Einstein's Relativity in Metaphor and Mathematics*, (New York: Prentice Hall).

Roboz, Elizabeth 1991, *Hans Albert Einstein*, (Iowa City: Iowa Institute of Hydraulic Research).

Romain, J. E., 1963, "Time Measurements in Accelerated Frames of Reference," *Reviews of Modern Physics*, 35 (April): 376-389.

Rosenkranz, Ze'ev, 1998 2002, *The Einstein Scrapbook*, (Baltimore, MD: Johns Hopkins University Press).

Rosenlicht, Maxwell, 1968 1986, *Introduction to Analysis*, (New York: Dover Publications).

Rosenthal- Schneider, Ilse, 1980, *Reality & Scientific Truth*, (Detroit: Wayne State University Press).

Ross, Sheldon M., 1980, *Introduction to Probability Models*, (New York: Academic Press).

Rosser, W. G. V., 1967, *Introductory Relativity*, (New York: Plenum Press).

Rosser, W. G. V., 1969, *Relativity & High Energy Physics*, (London: Wykeham Publications, Ltd.).

Rucker, Rudolf V. B., 1976 1977, *Geometry, Relativity, and the Fourth Dimension*, (New York: Dover Publications).

Rucker, Rudy, Editor, 1880 1980, *Speculations on the Fourth-Dimension*, Selected Writings of Charles H. Hinton, (New York: Dover Publications).

Rucker, Rudy, 1982 1983, *Infinity and the Mind: The Science and Philosophy of the Infinite*, (New York: Bantam Books).

Rucker, Rudy, 1984, *The Fourth Dimension: A Guided Tour of the Higher Universes*, (Boston: Houghton Mifflin Company).

Rucker, Rudy, 1985, *The Fourth Dimension: And How to Get There*, (London: Rider).

Rucker, Rudy, 2002, *Spaceland: A Novel of the Fourth Dimension*, (New York: Tom Doherty Book).

Russell, Bertrand, 1925 1956, *The ABC of Relativity*, (New York: The New American Library).

Sachs, Mendel, 1993, *Relativity in Our Time*, (London: Taylor & Francis).

Sachs, Robert, 1987, *The Physics of Time Reversal*, (Chicago: Chicago University Press).

Salmon, Wesley C., 1975, *Space, Time and Motion: A Philosophical Introduction*, (Encino, CA: Dickenson Publishing Co.).

Sartori, Leo, 1996, *Understanding Relativity*, (Berkeley, CA: University of California Press).

Savitt, S., Editor, 1995 1998, *Time's Arrows Today*, (Cambridge, England: Cambridge University Press).

Sayen, Jamie, 1985, *Einstein in America*, (New York: Crown Publishers, Inc.).

Schilpp, Paul Arthur, Editor, 1946, *The Philosophy of Bertrand Russell*, (Evanston, IL: Library of Living Philosophers, Inc.).

Schilpp, Paul, Editor, 1949, *Albert Einstein Philosopher-Scientist*, (Evanston, IL: Library of Living Philosophers, Inc.).

Schlick, Moritz, 1920, *Space and Time in Contemporary Physics*, (London: Oxford University Press).

Schlick, Moritz, 1949, *Philosophy of Nature*, (New York: Philosophical Library).

Schrodinger, Erwin, 1950, *Space-time Structure*, (Cambridge, England: Cambridge University Press).

Schumann, Charles, 1938, *Descriptive Geometry*, (New York: The Van Nostrom Company).

Schwartz, Jacob, 1962, *Relativity in Illustrations*, (New York: Dover Publications).

Schwartz, Joe, 1979, *Einstein for Beginners*, (New York: Pantheon Books).

Schwartz, Patricia and Schwartz, John, 2004, *Special Relativity: From Einstein to Strings*, (Cambridge, England: Cambridge University Press).

Schwinger, De Raad, Milton and Tsai, 1998, *Classical Electrodynamics*, (New York: Westview Press, Perseus Book Group).

Schwinger, Julian, 1986, *Einstein's Legacy*, (New York: Scientific American Library).

Sears, F. W. and Brehme, R. W., 1968, *Introduction to the Theory of Relativity*, (New York: Addison-Wesley Publishing Company).

Seeds, Michael A., 2001, *Foundations of Astronomy*, Sixth Edition, (New York: Brooks/ Cole).

Seelig, Carl, 1952, *Albert Einstein und die Schweiz*, (Zurich: Europa-Verlag).

Seelig, Carl, 1956, *Albert Einstein: A Documentary Biography*, (London: Staples Press Ltd.).

Seelig, Carl, Editor, 1956, *Helle Zeit - Dunkle Zeit*, (Zurich: Europa-Verlag).

Sellars, Roy Wood, 1926 1929, *The Principles and Problems of Philosophy*, (New York: The Macmillan Company).

Serviss, Garrett, 1923, *The Einstein Theory of Relativity*, (New York: Edwin Miles Fadmann, Inc.).

Shadowitz, Albert, 1968, *Special Relativity*, (New York: Dover Publications).

Shallis, Michael, 1983, *On Time*, (New York: Schocken Books).

Sheldon, H. Horton, 1932 1935, *Space, Time, and Relativity*, (New York: The University Society, Inc.).

Shlain, Leonard, 1991, *Art & Physics: Parallel Visions in Space, Time, and Light*, (New York: William Morrow).

Sierpinski, Waclaw, 1962 2003, *Pythagorean Triangles*, (New York: Dover Publications).

Sklar, Lawrence, 1974 1976, *Space, Time, and Spacetime*, (Berkeley, CA: University of California Press).

Sklar, Lawrence, 1985, *Philosophy and Spacetime Physics*, (Berkeley, CA: University of California Press).

Slosson, Edwin E., 1920, *Easy Lessons in Einstein*, (New York: Harcourt, Brace & Company).

Smith, David E., Editor, 1929, *Source Book in Mathematics*, (New York: McGraw-Hill Book Company).

Smith, James, 1965 1993, *Introduction to Special Relativity*, (New York: Dover Publications).

Smolin, Lee, 2006, *The Trouble with Physics*, (New York: Houghton Mifflin Company).

Smoot, G. and Davidson, K., 1993, *Wrinkles in Time*, (New York: Avon Books).

Sobel, Michael, 1987 1989, *Light*, (Chicago: University of Chicago Press).

Sommerville, D. M. Y., 1914 1958, *The Elements of Non-Euclidean Geometry*, (New York: Dover Publications).

Sommerville, D. M. Y., 1929 1958, *An Introduction to the Geometry of N Dimensions*, (New York: Dover Publications).

Sotheby's Catalog, 2002, *Geometry and Space*, April 2002, De Vitry Sale Catalog, (London: Sotheby's).

Sotheby's Catalog, 2004, *Masterpieces from the Time Museum, Part Four*, Three Volumes, (New York: Sotheby's).

Specker, Hans Eugen, Editor, 1979, *Einstein and Ulm*, (Ulm & Stuttgart: Kommissionsverlag W. Kohlhammer).

Spector, Marshall, 1972, *Methodological Foundations of Relativistic Mechanics*, (Notre Dame, IN: University of Notre Dame Press).

Speyer, Edward, 1994, *Six Roads from Newton*, (New York: John Wiley & Sons, Inc.).

Spinoza, Baruch, 1663 1961, *Principles of Cartesian Philosophy*, (New York: Philosophical Library).

Stachel, John, Editor, 1987, *Collected Papers of Albert Einstein*, Volume 1, The Early Years (1879-1902), (Princeton: Princeton University Press).

Stachel, John, Editor, 1998, *Einstein's Miraculous Year*, (Princeton: Princeton University Press).

Stachel, John, 2002, *Einstein from B to Z*, (New York: Birkhauser).

Stanley, Thomas, 1687 1970, *Pythagoras*, Reprint of the History of Philosophy, (Los Angeles, CA: Philosophical Research Society, Inc.).

Steiner, Rudolf, 1995 2001, *The Fourth Dimension: Sacred Geometry, Alchemy and Mathematics*, (Great Barrington: Anthroposophic Press).

Steinmetz, Charles P., 1923, *Four Lectures on Relativity and Space*, (New York: McGraw-Hill Book Company).

Stephens, Carlene E., 2002, *On Time: How America Has Learned to Live by the Clock*, (Boston: Little, Brown and Company).

Stern, Fritz, 1999, *Einstein's German World*, (Princeton: Princeton University Press).

Stewart, Ian, 2001, *Flatterland*, (Cambridge, M A: Perseus).

Straumann, Norbert, 1936 1991, *General Relativity and Relativistic Astrophysics*, (Berlin, New York: Springer-Verlag).

Strawson, P. F., 1966, *The Bounds of Sense: An Essay on Kant's Critique of Pure Reason*, (London: Methuen & Co. Ltd.).

Sullivan, Walter, 1979, *Black Holes: The Edge of Space, the End of Time*, (Garden City, N Y: Anchor Press/ Doubleday).

Synge, J. L., 1960, *Relativity: The General Theory*, (Amsterdam: North Holland Publications).

Szu-Whitney and Whitney, 1999, *Portals & Corridors: A Visionary Guide to Hyperspace*, (Berkeley, CA: Frog, Ltd.).

Tabak, John, 2004, *Geometry: The Language of Space and Form*, (New York: Facts on File, Inc.).

Tauber, Gerald E., 1979, *Man and the Cosmos*, (New York: Greenwich House, Crown Publishers).

Tauber, Gerrald, Editor, 1979, *Albert Einstein's Theory of General Relativity*, (New York: Crown Publishers).

Taylor, E. and Wheeler, J., 1963, *Spacetime Physics*, First Edition, (New York: W. H. Freeman).

Taylor, E. and Wheeler, J., 1992, *Spacetime Physics*, Second Edition, (New York: W. H. Freeman and Company).

Thorne, Kip, 1994, *Black Holes & Time Warps*, (New York: W. W. Norton & Company).

Toben, B. and Wolf, F., 1975 1987, *Space-time and Beyond*, (New York: Bantam Books).

Tolman, Richard, 1934 1987, *Relativity, Thermodynamics and Cosmology*, (New York: Dover Publications).

Tonnelat, Marie-Antoinette, 1959 1966, *The Principles of Electromagnetic Theory and of Relativity*, (Netherlands: D. Reidel Publishing).

Tornebohm, Hakan, 1963, *Concepts and Principles in the Space-time Theory Within Einstein's Special Theory of Relativity*, (Gothenburg: Gothenburg Studies in Philosophy).

Torretti, Roberto, 1983 1996, *Relativity and Geometry*, (New York: Dover Publications).

Trbuhovic-Gjuric, Desanka, 1988, *Im Schatten Albert Einstein*, (Bern and Stuttgart: Paul Haupt).

Tribble, A., 1996, *Princeton Guide to Advanced Physics*, (Princeton: Princeton University Press).

Troop, Frenkel, and Chernin, 1993, *Alexander A. Friedmann: The Man Who Made The Universe Expand*, (Cambridge: Cambridge University Press).

Tulka, Tarthang, 1977, *Time, Space, & Knowledge: A New Vision of Reality*, (Berkeley, C A: Dharma Publishing).

Turner, A. J., Editor, 1990, *Time, Exhibition Catalog, the Hague*, (Amsterdam: The Hague).

Tyndall, John FRS, 1897, *Fragments of Science*, Sixth Edition, Volume 1, (New York: Appleton and Company).

Tyndall, John FRS, 1897, *Fragments of Science*, Sixth Edition, Volume 2, (New York: Appleton and Company).

Vallentin, Antonina, 1954, *Drama of Albert Einstein*, (New York: Doubleday & Company).

Van Fraaseen, Bas C., 1970, *An Introduction to the Philosophy of Time and Space*, (New York: Random House).

Van Heel & Vezel, 1968, *What Is Light?*, (New York: McGraw Hill).

Vasiliev, A. V., 1924, *Space Time Motion*, (New York: Alfred A. Knoff).

Vladimirov, et. Al., 1984 1987, *Space Time Gravitation*, (Russia: Mir Publishers).

Wald, Robert, 1977, *Space, Time, and Gravity*, (Chicago: University of Chicago Press).

Wald, Robert, 1984, *General Relativity*, (Chicago: University of Chicago Press).

Wald, Robert, Editor, 1998, *Black Holes and Relativistic Stars*, (Chicago: University of Chicago Press).

Waugh, Alexander, 1999, *Time: From Micro-seconds to Millennia - A Search for the Right Time*, (London: Headline Book Publishing).

Weitzenbock, Roland, 1956, *Der Vierdimensionale Raum*, (Stuttgart: Berkhauser Verlag Basel).

Weyl, Hermann, 1920 1952, *Space-Time-Matter*, Fourth Edition, (New York: Dover Publications).

Wheeler and Ciufolini, 1995, *Gravitation and Inertia*, (Princeton: Princeton University Press).

Wheeler, John, 1990, *A Journey Into Gravity and Spacetime*, (New York: Scientific American Library).

White, L. L., 1931, *Critique of Physics*, (London: Kegan Paul, Trench, Trubner & Co. Ltd.).

White, M. and Gribbin, J., 1992, *Stephen Hawking: A Life in Science*, (New York: Penguin Books).

White, M. and Gribbin, J., 1993 1995, *Einstein a Life in Science*, (New York: Penguin Books).

Whitehead, Alfred North, 1925 1967, *Science and the Modern World*, (New York: Free Press Paperback).

Whitrow, G. J., 1955, "Why Physical Space Has Three Dimensions," The British Journal for the Philosophy of Science, 6 (May): 13-31.

Whitrow, G. J., 1967 1973, *Einstein the Man and His Achievement*, (New York: Dover Publications).

Whitrow, G. J., 1980, *The Natural Philosophy of Time*, Second Edition, (Oxford: Claredon Press Oxford).

Whitrow, G. J., 1988, *Time in History*, (Oxford: Oxford University Press).

Whitrow, G. J., Editor, 1967, *Einstein: The Man and His Achievement*, (New York: Dover Publications).

Wiborg, James H., 1992, *Theoria Primarius: On Force, Time, and Observer in Space-Time*, (Tacoma, WA: Published by Author).

Wiener, P. and Noland, A., Editors, 1957, *Roots of Scientific Thought*, (New York: Basic Books).

Wilczek, F. and Devine, B., 1987 1989, *Longing for the Harmonies: Themes and Variations from Modern Physics*, (New York: Norton & Co.).

Will, Clifford, 1981, *Theory and Experiment in Gravitational Physics*, (Cambridge: Cambridge University Press).

Will, Clifford, 1986, *Was Einstein Right?*, (New York: Basic Books).

Wilson, Jerry D., 1990, *College Physics*, Second Edition, (New York: Prentice Hall).

Wittaker, Sir Edmund, 1910 1951, *History of the Theories of Aether and Electricity*, Volume 1 Classical, Volume 2 1900-1926, (London: Thomas Nelson & Sons Ltd.).

Wittgenstein, Ludwig, 1932 1978, *Philosophical Grammar: Part 1 The Proposition and its Sense, Part 2 On Logic & Math*, (Berkeley: University of California Press).

Wittgenstein, Ludwig, 1956, *Remarks on the Foundation of Mathematics*, (Oxford: Oxford University Press).

Wolf, Fred A., 1988, *Parallel Universes*, (New York: Simon & Schuster).

Wolf, Fred, 1989, *Taking the Quantum Leap, the New Physics for Non-Scientists*, (New York: Harper & Row, Publishers).

Wolfson, Richard, 2000, *Einstein's Relativity and the Quantum Revolution*, Video Lecture Series, (Chantilly, VA: The Teaching Company).

Wolfson, Richard, 2003, *Simply Einstein*, (New York: Norton & Co.).

Woolf, Harry, Editor, 1980, *Some Strangeness in the Proportion: A Centennial Symposium*, (London: Addison-Wesley Publishers).

Wright, Lawrence, 1992, *Clockwork Man: The Story of Time, its Origins, its Uses, its Tyranny*, (New York: Barnes & Noble).

Young, Louise, Editor, 1963, *Exploring the Universe*, (New York: McGraw-Hill Book Company, Inc.).

Young, Nicholas, 1988 1992, *An Introduction to Hilbert Space*, (New York: Cambridge University Press).

Yourgrau, Palle, 2005, *A World without Time: The Forgotten Legacy of Godel and Einstein*, (New York: Basic Books).

Zajonc, Arthur, 1993, *Catching the Light*, (Oxford: Oxford University Press).

Zee, A., 1989, *An Old Man's Toy*, (New York: Macmillan Publishing Company).

Zeilik, Michael, 2002, *Astronomy the Evolving Universe*, Ninth Edition, Textbook, (Cambridge, England: Cambridge University Press).

Zeldovich, Y. and Novikov, I., 1971 1996, *Stars and Relativity*, (New York: Dover Publications).

Zeldovich, Y. and Novikov, I., 1983, *The Structure and Evolution of the Universe*, (Chicago: University of Chicago Press).

Zippin, Leo, 1962 2000, *Uses of Infinity*, (New York: Dover Publications).

Index

Abbott 18, 57

absolute motion 42

absolute time 8, 15

accelerated motion 67, 78, 95

Alexander 12, 15, 59

arrow of time 71

astronomy 68

    black hole 75

before Einstein 18

Begley 68

black hole 23, 72

    escape from a 76

    falling into a 73, 74

Boslough 70

Broad 15, 17, 86

Bunch 70

Burger 18, 57

Burtt 16

Callahan 16

Campbell 16

Capra 35, 90

confusing ideas

    black hole 73

    Einstein's first paper 5

    light cone 6

    relative timekeeping 8

    space contracts 4

    spacelike 7, 22

    time dilation 1

timelike 22

triplet paradox 5

twin acceleration 3

twin paradox 2

warped space 61

who's moving? 3, 22

Coveney 70

Davies 68, 70, 71

definition for time 9, 32

Descartes 27, 28

Doppler shift 37, 46

D'Abro 16

Eddington 8, 12, 38

eigenzeit 8

Einstein's ideas

    equivalence principle 92

    four-dimensional field 95

    general relativity 78, 89

    invariance 84

    no absolute motion 86

    no absolute rest 35, 36

    no empty space 12, 93

    non-Euclidean geometry . . 59

    observing 86, 87

    principle of equivalence . . 41

    relative motion . . . 36, 37, 86

    relative speed 49

    relative timekeeping . . . 8, 38

    special theory 12, 51, 78

    speed of light 50, 54

    time dilation 1, 54

    warped space 61

Ellis 57
entropy 71
Epstein 29

Euclidean geometry
. . . . 11, 20, 60, 96
    fourth dimension 60
    limitation 79
    solution 56, 77, 79, 95
    timekeeping rate 47
    Timespace 78
    vision 58

Feynman 71
FitzGerald 49
fly inside automobile 39
fly unaware of motion 43
fly outside of auto 49
four-dimensional field 95

fourth dimension 18, 57
    definition 55
    hypercube 63, 64
    speed of light 26
    time dilation 55

Frank 17
Gardner 70
Gedankin experiment . 23, 59, 65

general relativity 88, 95
    time's variable rate 91

Giancoli 43
Gödel 2

Gott 68
Greene 13, 29, 41
Gross 9
Halsted 96
Hawking 29, 71
higher dimensions 61, 65
higher mathematics 89
Hilbert 58, 80, 84
Hinton 18, 57
hypercube 63

invariance
    diameter of a circle 80
    Euclidean geometry 84
    metric 16
    non-Euclidean geometry 81, 82
    speed of light 16

Jammer 12, 13
Kaku 13, 62, 68
Klein 70
Krauss 13
lab time 8
Leibniz-Clarke 12
length contraction 4
lightlike 7
Lobachevski 95
local time 8

Lorentz 8
    contraction 50, 51

Michelson-Morley 49, 51
Miller 12, 37, 84

Minkowski
biography 19
space-time 21

Minkowski's ideas
light cone 6
Raum und Zeit 19, 21
space-time 6, 19, 84
space-time continuum . 11, 22
spacelike 7
timelike 7

Moore 17

motion in space 14, 21, 71
absolute motion 41, 42
accelerated motion 67
arrow of time 71
fourth dimension 72
no absolute motion 42
speed of light .. 26, 27, 29, 55
Timespace 52
unaware of 39, 40

Murchie 84
natural philosophy 14
need better ideas 9, 18

New ideas 9, 58
Universal Time 8
black hole 75
empty space exists 93
event horizon 74
fourth dimension 3
motion in four dimensions 11

physical space 3
time reversal 74, 75
timekeeping's variable rate
.... 33, 46
Timespace 24, 33, 74
Universal Equivalence
Principle 41, 92
Universal Time 24, 31,
33, 74, 75
Universal Timekeeping
Frequency 46

Newton's ideas
absolute time 3, 8, 31
criticism of 15
force 88
no absolute motion 15
universal gravity 45

non-Euclidean geometry
.... 11, 20, 79
parallel theory 16
Penrose 29

philosophy
eastern 35
natural 14
pure 15, 16

physical space 3, 11, 14, 25
four dimensions 64

Pias 90

Poincaré 12
    four dimensions 19
projections 62, 64, 65
proper time 8
Pythagorean Theorem 27, 28
Randall 13
Rees 12, 13, 88
Reichenbach 68, 71
relative motion 87
relative speed 49
Riemann 20
Rucker 58, 68
Sachs 70, 71
Salmon 18
Savitt 70

scalar 25
    scalar time 34, 36
    time 26

Schwarzschild radius 73, 76
Seeds 72

seeing 60, 67
    four dimensions 57
    twin paradox 66

Shallis 70
singularity 72, 73
Sklar 68, 70, 71
Smolin 9, 13
Sommerville 95

space
    absolute 17
    contraction 51
    definition 9
    dimension 10
    four dimensions 12, 17
    higher dimensions 61
    physical space 3, 11, 22
    relative 17
    science of space 12
    three dimensions . 12, 17, 18
    warped 67
    warped space 90

space and time 11, 16
space-time 15, 17
spacelike 7
spaceships A and B 50
special theory 12, 78
Steinmetz 21, 22
Stewart 58
superstring theory 65, 93
Taylor 7, 8, 80, 84
Thorne 71

three dimensions 12, 22, 72
    common notions 23

time
    common notions 23
    definition 9, 32, 38
    definition for time 40
    eigenzeit 8
    lab time 8
    local time 8

time (continued)
    physical direction 25
    proper time 8
    scalar time 42
    time dilation 1

    timekeeping rate 91
    time's variable rate . . . 45, 47
    unaware of motion 43
    with mathematical truth 2

time reversal 68-71

time travel 68
    aging 69

timekeeping frequency 34, 36
timekeeping rate 91
timelike 7

Timespace
    . . . . . 29, 30, 33, 38, 77-79,
              85, 87, 88, 95
    aging 52
    contraction 51, 52
    definition 30
    displacement 60
    negative displacement 29
    time dilation 53
    vector displacement . . . 26, 28
    velocity 30

triplet paradox 5
twin paradox 2, 8, 26, 65-69

Universal Equivalence Principle
    . . . . 92, 93

Universal Reference Frame
    . . . . 40, 41, 53, 56,
             86, 87, 92, 93
    definition 43
    motion 93

Universal Time 30, 31,
    36, 39-41, 47, 53, 56,
    77, 79, 85, 87, 92, 95
    examples 44
    measurement 48

van Fraassen 17
Vasiliev 95
vector 25
warped space . . 61, 62, 67, 89, 90
Wheeler 7, 8, 80, 84
Whitrow 17, 68, 70
Whyte 16
Will 68, 91
Wilson 4
world Line 7
Yourgrau 2

Author

Thomas W. Sills is a professor at Wright College, one of the City Colleges of Chicago. He worked on course and lab development at several teaching positions over forty years.

His diverse professional career in science education includes writing, psychological test development, science toy design, science teacher training, reviews of college physics textbooks, and consultant to science programs for the gifted. He has also developed educational courses on Channel 20/Chicago television, including *The Mechanical Universe* and *Planet Earth*.

In high school at the International Science Fair he received an award for his science project on learning and memory. In 1967 his first college teaching assignment was physical science for elementary teachers. In 1977 he received a Ph.D. in both physics and education at Purdue University.

Dr. Sills is a serious collector of books and manuscripts on science and technology. But most of all, he enjoys going to new places to meet new people in order to stimulate ideas. "Doing adventuresome things is just plain fun."
